T0298341

Introduction to Earthquake Engineering

Introduction to Earthquake Engineering

Hector Estrada and Luke S. Lee

CRC Press

Taylor & Francis Group

Boca Raton London New York

CRC Press is an imprint of the
Taylor & Francis Group, an **informa** business

CRC Press
Taylor & Francis Group
6000 Broken Sound Parkway NW, Suite 300
Boca Raton, FL 33487-2742

© 2017 by Taylor & Francis Group, LLC
CRC Press is an imprint of Taylor & Francis Group, an Informa business

No claim to original U.S. Government works

Printed on acid-free paper

International Standard Book Number-13: 978-1-4987-5826-0 (Hardback)

Library of Congress Cataloging-in-Publication Data

Names: Estrada, Hector, author. | Lee, Luke S., author.
Title: Introduction to earthquake engineering / Hector Estrada and Luke S. Lee.
Description: Boca Raton : Taylor & Francis, CRC Press, 2017. | Includes bibliographical references and index.
Identifiers: LCCN 2017018885| ISBN 9781498758260 (hardback : acid-free paper) | ISBN 9781315171579 (ebook)
Subjects: LCSH: Earthquake engineering.
Classification: LCC TA654.6 .E848 2017 | DDC 624.1/762--dc23
LC record available at https://lccn.loc.gov/2017018885

Visit the Taylor & Francis Web site at
http://www.taylorandfrancis.com

and the CRC Press Web site at
http://www.crcpress.com

Contents

Preface

Earthquake engineering is an important part of structural engineering. The primary purpose of *Introduction to Earthquake Engineering* is to provide students and practitioners with the fundamental and practical knowledge of how earthquake engineering is used in the design of lateral force-resisting systems for buildings and other structures. Special emphasis is placed on obtaining seismic loads using the most recent code specifications.

The topics and illustrations presented in this text have been carefully chosen for students in courses dealing with seismic loads as part of an architectural, construction, or civil engineering-related curriculum: both at the undergraduate and graduate levels. An important and unique feature of *Introduction to Earthquake Engineering* is that it ties together the relationship and interdependence between structural dynamics and seismic loads based on code specifications. With this in mind, many practical examples are provided throughout the text. Many solutions to example problems are prepared using computer programing methods (i.e., MATLAB®) to enable students, and practitioners, to solve complex dynamics problems. Also, considerable content is available to acquaint students with specifications and reference information found in the American Society of Civil Engineers' *Minimum Design Loads for Buildings and Other Structures*.

Examples of typical seismic load calculations presented in this text demonstrate the type of routine technical mathematics an entry-level structural engineer can expect to encounter on the job. Because *Introduction to Earthquake Engineering* is first and foremost an introductory textbook, most complex nonlinear analysis theory has been purposely avoided. However, the textbook does assume that the reader has had some exposure to rigid body dynamics, differential equations, and structural analysis coursework. The 2010 edition of the *Minimum Design Loads for Buildings and Other Structures* (ASCE-7) is the source of all design code-based specifications.

This book provides broad coverage of the subject of earthquake engineering, emphasizing the process of establishing seismic loads for buildings and other structures. To that end, the reader is introduced to a wide variety of practical dynamics and seismic loading examples and end-of-chapter problems. Also, to guide the reader, each chapter begins with a list of learning outcomes. After carefully studying this textbook, a reader should be able to: (i) demonstrate a basic understanding of earthquake engineering, (ii) identify and solve fundamental structural dynamics problems, and (iii) perform equivalent static and dynamic seismic analyses of simple structural systems. The first objective relates to the student's ability to operate in the first two cognitive domains of Bloom's Taxonomy (viz., the knowledge of seismic engineering terms and the comprehension of the overall area of earthquake engineering). The second and third objectives primarily concentrate on the next two cognitive domains (viz., the application of structural dynamics in earthquake engineering and the analysis of systems subjected to earthquake loading).

Image galleries, PowerPoint presentations, a solutions manual, etc., will be hosted on a companion website. Visit the book's CRC Press website for further details:

https://www.crcpress.com/Introduction-to-Earthquake-Engineering/Estrada-Lee/p/book/9781498758260

Hector Estrada
Luke S. Lee

MATLAB® is a registered trademark of The MathWorks, Inc. For product information, please contact:

The MathWorks, Inc.
3 Apple Hill Drive
Natick, MA 01760-2098 USA
Tel: 508-647-7000
Fax: 508-647-7001
E-mail: info@mathworks.com
Web: www.mathworks.com

Acknowledgements

We are grateful to the many students at University of the Pacific whose suggestions and reactions to our earthquake engineering classes have driven the need for us to refine our presentation of earthquake engineering as simply and clearly as possible. Finally, we are indebted to Wendy Estrada for her proofreading and extensive checking of details: both technical and stylistic, and most importantly for all of her exhaustive support during the writing process.

Authors

Hector Estrada, PhD, PE, is currently a professor of civil engineering at University of the Pacific (Pacific), where he also served as chair of the Civil Engineering Department. Prior to joining Pacific, Dr. Estrada was chair of the Department of Civil and Architectural Engineering at Texas A&M University-Kingsville. Dr. Estrada has also had visiting research appointments at NASA, Texas Tech University, and the U.S. Army Engineer Research and Development Center— Construction Engineering Research Laboratory. His teaching interests include structural engineering and mechanics, the design of concrete, steel, and timber structures, structural dynamics, and earthquake engineering. He has published on structural engineering and engineering education in various peer-reviewed journals, conference proceedings, and presented research work at various technical conferences. He has published a book on drafting and design of structural steel buildings, and several book chapters. He has served as a reviewer for a number of journals, conferences, book publishers, and funding agencies. Dr. Estrada earned his BS (with honors), MS, and PhD all in Civil Engineering from the University of Illinois at Urbana-Champaign.

Luke S. Lee, PhD, PE, is currently an associate professor of civil engineering at University of the Pacific, where he teaches courses in solid mechanics, structural dynamics and health monitoring, structural design, and engineering risk analysis. He has authored and co-authored numerous journal articles, conference papers, and book chapters in composite materials, structural health monitoring, service life estimation, and engineering pedagogy. Dr. Lee earned his BS in civil engineering from the University of California, Los Angeles, MS in civil engineering from the University of California, Berkeley, and PhD in structural engineering from the University of California, San Diego.

1 Introduction

After reading this chapter, you will be able to:

1. Explain the general effects of earthquakes on the built environment
2. Describe the main source of earthquakes
3. List the different types of earthquakes
4. Explain tectonic plate movement

"If a tree falls in a forest and no one is around to hear it, does it make a sound?" Since sound is the sensation excited in the ear when air is set in motion, one could argue that the tree makes no sound. Likewise, if an earthquake occurs and no one, or no infrastructure system, is there to experience the resulting motion, we can conclude that the most severe earthquake would cause no damage. Before humans appeared on earth, numerous earthquakes occurred, as evidenced by the large shift in the location of the continents with no damage (this will be discussed in more detail in Section 1.2.3). So we could argue that earthquakes do not cause damage; rather, it is infrastructure that experiences damage because of an earthquake. That is, the devastating effect of earthquakes should be attributed to an inferior built environment. Thus, when infrastructure is designed to account for the effects of earthquakes, even the largest earthquakes cause little or no damage! Nevertheless, because of the historical effects of earthquakes on human civilization, they are the most terrifying of all natural disasters. The principle reason being that they can occur with little or no notice and can devastate large areas.

This book is an introduction to the essentials of seismic design. The primary focus is on establishing the forces and deformations experienced by structural systems found in infrastructure. We begin by providing a broad overview of earthquakes, their effects on structures, sources, and types (Chapter 1), followed by an introduction to seismology in Chapter 2, the material for which was selected based on seismology material covered in the California seismic portion of the Professional Engineering (PE) exam. The book then covers the prerequisite structural dynamics material (based on single-degree-of-freedom [SDOF] dynamic analyses) needed to develop the fundamental relationships between structures and ground motion produced by earthquakes (Chapter 3). The material being primarily focused on the development of parameters that are based on free and forced, damped and un-damped vibrations, with the aim of developing parameters to characterize structural system mass, stiffness, and damping, which in turn are used to calculate periods, frequencies, and other relevant vibration properties.

We also explain the computational processes used to develop a critical concept (Chapter 4), which is the establishment of seismic forces for structures at a particular location: the response spectrum based on the Newmark–Hall method—both elastic and inelastic spectra (Chapter 5). Next, we connect the analysis of multi-degree-of-freedom (MDOF) building systems (or multistory shear buildings) to fundamental SDOF principles through the development of the generalized SDOF equations (Chapter 6). This generalized SDOF formulation can be combined with the principles associated with SDOF analyses, including response spectrum analysis, to analyze a multistory building as an SDOF system, including finding maximum shear and bending moment in all members due to a seismic load. The process entails computing approximate participation factors and using a design response spectrum to determine the story displacements, which are then used to obtain lateral story seismic forces that are necessary to perform a complete analysis for the internal loading of a structural system.

Following the response spectrum analysis of MDOF systems subjected to seismic ground motion in Chapter 7, we examine code-based analyses—both equivalent-static and dynamic approaches in Chapter 8. The dynamic analysis will emphasize modal response spectrum analysis. The main goal of the book is to elucidate code-based formulations such that a practicing engineer or an engineering student can apply these formulations, and explain how they were developed and describe the limitations of these code-based procedures.

The book is intended as a tool for a one-semester course in introduction to earthquake engineering at the undergraduate or graduate levels, both in civil or architectural engineering.

The material presented in the book is concise and includes both essential structural dynamics and seismic engineering principles, the former being prerequisite to fully grasping the fundamental concepts covered in the latter. Thus, students who have a limited background in structural dynamics still have the opportunity to cover the essentials of earthquake engineering in a one-semester course. This is particularly important in an era where information is readily available and can be overwhelming for a new learner. For completeness, we also include computational tools that can be used to solve the complex mathematics involved in the solution of even the simplest seismic engineering problems.

At the beginning of each chapter, we present a list of learning outcomes in an effort to improve the delivery of the content and to guide the learner in mastering the material presented. Overall, the book adheres to the following general learning outcomes:

1. Demonstrate a basic understanding of earthquake engineering
2. Identify and solve basic structural dynamics problems
3. Perform basic equivalent static and dynamic seismic analyses of simple structural systems

The first objective relates to the student's ability to operate in the first two cognitive domains of Bloom's Taxonomy (namely, the knowledge of seismic engineering terms and comprehension of the overall area of earthquake engineering). The second and third objectives primarily concentrate on the next two cognitive domains (namely, the application of structural dynamics in earthquake engineering and the analysis of systems subjected to earthquake loading).

1.1 STRUCTURAL EFFECTS OF EARTHQUAKES

Generally, effects of earthquakes can be divided into several categories. In this section, we present three categories: ground failure, indirect effects, and ground shaking, the latter being the main topic covered in the book. Ground failure can be further subdivided into surface rupture, ground subsidence, ground cracking, soil liquefaction, and landslides. Indirect effects include tsunamis (tidal waves in the oceans), seiches (tidal waves in an enclosed body of water, such as a lake), and fires, all of which can result in damage comparable to, or even larger than, ground shaking.

1.1.1 GROUND FAILURES

Ground failures are generally considered part of geotechnical earthquake engineering, and they involve the movement of the ground surface at a location where geological fissures or zones of weakness in the crust of the earth (faults) slip slowly or suddenly. Usually, sudden movement is much more damaging because of the associated ground shaking. The terminology and the process of how earthquakes occur will be discussed in more detail in the rest of this chapter. In this section, we present an overview of the most severe effects of ground failure on infrastructure.

Surface faulting occurs when the relative movement of rocks on the two sides of a fault takes place deep within the earth and breaks through to the surface; this can occur as slow movement in the form of *fault creep* or suddenly resulting in an *earthquake*. This type of ground failure typically follows a preexisting fault line. Figure 1.1a shows the slow movement of the San Andreas Fault inside the DeRose Winery in the Cienega Valley near Hollister, California. This location

(a)   (b)

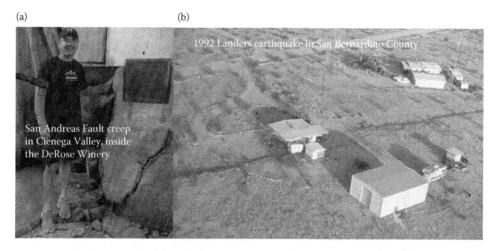

1992 Landers earthquake in San Bernardino County

San Andreas Fault creep
in Cienega Valley, inside
the DeRose Winery

FIGURE 1.1   Surface rupture: (a) slow (creep) and (b) caused by the 1992 Landers earthquake in San Bernardino County. (http://www.conservation.ca.gov/cgs/rghm/ap/Pages/main.aspx)

experienced over 16 inches (400 mm) of fault creep over a 40-year period as indicated by a graph of fault slip displayed at the winery. Since the winery building straddles the surface of the fault, this movement has been splitting it into two, which is clearly evident by the large crack on the concrete floor. The San Andreas Fault, however, is capable of producing large earthquakes near this site, such as the 1989 Loma Prieta earthquake, which caused widespread damage throughout the San Francisco Bay Area, but did not break through to the surface. Figure 1.1b shows the surface rupture resulting from the 1992 Landers earthquake, in San Bernardino County, California. The sudden slip of the fault extended for nearly 50 miles with relative displacements from 1 inch to 20 feet.

Ground subsidence occurs as loose soils rearrange and settle into a denser state during vibrations caused by earthquakes. In some cases, the compaction effect may amount to substantial settlement of the ground surface. Liquefaction, which will be discussed later in this section, can also trigger significant subsidence of unconsolidated sediment. For example, it is believed that Big Lake and St. Francis Lake in Northeast Arkansas formed during the 1811 and 1812 New Madrid earthquakes in areas where liquefaction resulted in significant lowering of the land surface, causing it to flood. Uniform settlement over large tracks of land causes few problems for infrastructure (except for long horizontal structures, such as roads and pipelines). However, systems along the boundaries can experience differential settlement that can be extremely damaging, as was experienced in Anchorage, Alaska along Fourth Avenue during the Good Friday earthquake in 1964; see Figure 1.2.

Ground cracking is usually observed along the edges of ground subsidence as shown in Figure 1.2; it may also be the result of slope failure or liquefaction, all of which cause the ground to lose its support and sink, with the ground surface breaking up into *fissures, scarps, horsts,* and *grabens* (Figure 1.2). The most damaging effect of ground subsidence is differential settlement, which can severely disrupt the function of any infrastructure system near the vicinity of cracking locations, particularly those with long foundations that straddle the cracks.

Soil liquefaction occurs when loose, saturated granular soils temporarily change from a solid to a liquid state, losing their shear strength, which corresponds to a loss in effective stress between soil particles. Loose saturated (or moderately saturated) sands and nonplastic silts are most susceptible to this ground failure; however, in rare cases, gravel and clay can also experience liquefaction. In all cases, poor drainage within the loose soil causes an increase in the pore water pressure as the soil is compressed by the vibratory effect of seismic waves. As the load is transferred from soil to pore water pressure, the effective stress between particles is temporarily reduced, or eliminated, causing

FIGURE 1.2   Surface subsidence. (Courtesy of USGS.)

a corresponding decrease in shear strength. In some cases, the pore water pressure increases rapidly and a slurry (soil water mixture) forms that flows vertically to the surface, which can result in craters and sand boils as shown in Figure 1.3a. Liquefaction is responsible for some of the most spectacular failures caused by earthquakes as shown in Figure 1.3b. Following the 1964 Niigata, Japan earthquake, several apartment buildings experienced severe tilting and settlement. Fortuitously, the tilted buildings did not suffer major structural damage, and thus minimal human injury.

Landslides caused by earthquakes are uncommon. Consequently, in order for a structure to experience damage during the event, it must be located at the top or bottom of the soil mass that slides down; for this reason, damage resulting from earthquake-induced landslides is rare. Sloped land that is marginally stable under static conditions is most susceptible to sliding during the intense shaking of strong earthquakes. For the most severe cases, debris (soil, boulders, and other materials) flow can move at avalanche speeds and can travel long distances depending on the slope from which the landslide was formed. Furthermore, earthquake-induced landslides can be sudden

(a)   (b)

FIGURE 1.3   (a) Sand boils that erupted following the 2011 Christchurch Earthquake and (b) Tilting and settlement of buildings caused by liquefaction following the 1964 Niigata Earthquake. (Courtesy of USGS.)

(a)   (b)

Vista parcial de plaza mayor de la ciudad de Yungay, Peru 1966   Foto: Aurelio Alva Mendez

FIGURE 1.4   (a) Partial view of Yungay City, Peru in 1966 and (b) after being buried by the 1970 Great Peruvian earthquake-induced landslide. (Courtesy of EERC, University of California.)

and unpredictable, producing the total destruction of communities in the path of the debris flow. Some devastating landslides have occurred around the world. One of the worst cases happened in Peru during the 1970 Ancash (or Great Peruvian) earthquake (magnitude 7.9) off the Pacific Ocean coast, which produced what is considered the world's deadliest earthquake-induced landslide (20,000 fatalities). During the earthquake, the northern wall of Mount Huascaran, 130 km (70 miles) from the epicenter, was loosened, mobilizing rock, ice, and snow into a massive landslide that buried the towns of Ranrahirca and Yongay; see Figure 1.4.

1.1.2 INDIRECT EFFECTS OF EARTHQUAKES

These include some of the most devastating and frightening impacts of earthquakes, such as tsunamis and fires. Although these cause major damage to structural systems, they are only briefly discussed in this book. Designing structural systems to withstand the large magnitude forces resulting from water waves generated by a tsunami is challenging. Tsunamis are long-period sea waves that are generated when an earthquake causes the vertical movement of the seafloor. Tsunamis travel far, at high speeds (over 500 mph) in the open ocean and are difficult to detect because of their small crest-to-trough height, and long wavelengths, which typically, are hundreds of miles long. Unobstructed, these waves can travel around the world and dissipate all their energy without causing damage. However, as they approach a shore, the water depth decreases causing an increase in wave speed and wave amplitude (height of wave run-ups). Wave run-ups of 75 feet have been observed at several locations. Wave run-ups can push water that rushes far inland, and have created devastating damage to infrastructure and great loss of life. One of the most devastating tsunami events occurred in the Indian Ocean on December 26, 2004, which killed over 230,000 people in 14 different countries; the furthest fatality was in Port Elizabeth, South Africa, over 8000 km away (see Figure 1.5). In some flat areas, such as Aceh on the northwest tip of Sumatra Island in Indonesia, the wave run-ups pushed water as far as 2 km inland, with wave heights of over 25 m. The wave forces generated by these types of events are nearly impossible to design against.

Seiches are earthquake-induced waves in an enclosed body of water, such as a lake or a reservoir, or one that is partially enclosed, such as a bay. Seiches are caused when long-period seismic waves resonate with oscillations of the enclosed water and cause standing waves. Earthquakes may happen within or far outside the perimeter of the body of water. Although this type of wave has been observed during most earthquakes (even in swimming pools), related damage to infrastructure has been minimal.

Fire is probably the most terrifying indirect effect of earthquakes, particularly considering that people who survived in collapsed buildings, but were trapped in the debris, were burnt alive

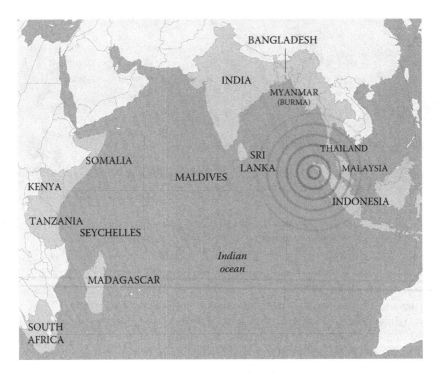

FIGURE 1.5   Countries affected by the 2004 Indian Ocean Tsunami. (From Wikipedia.)

as the fire consumed everything in its path. Traditional firefighting methods are often ineffective against earthquake-induced fires because most water mains that supply water hydrants break. Earthquake-induced fires are started by ruptured combustible substance conduits (such as gas mains) or destroyed combustible substance storage containers (such as oil tanks), and then ignited by sparks from sources such as downed powerlines. Earthquake-induced fires are most common in regions where timber construction is prevalent. One of the most devastating earthquake-induced fires happened after the 1906 San Francisco earthquake. Over a four-day period, the fire burned nearly 25,000 buildings on 508 city blocks (see Figure 1.6), which constituted about 80% of the city at the time.

FIGURE 1.6   Extent of San Francisco fire following the 1906 Earthquake. (Courtesy of USGS.)

1.1.3 GROUND SHAKING

Ground shaking causes the majority of earthquake damage; additionally most of the aforementioned effects are caused by shaking. In fact, where the shaking intensity is low, the hazard of other effects can be minimal. Consequently, shaking is the only effect experienced by everyone within an afflicted area, and intense shaking can produce widespread damage from various seismic hazards. For this reason, ground shaking is the main focus of earthquake engineering.

Although earthquakes are caused by numerous natural and human-induced phenomena, the events posing the highest seismic risks are caused by the relative deformation of crustal tectonic plates (discussed in Section 1.2.3). Most frequently, earthquakes happen at the boundaries, but in rare occasions they occur far from the boundaries. The most spectacular examples of these are the aforementioned New Madrid earthquakes. The boundary interface of adjacent plates is generally rough, with friction resisting the relative deformation of these plates. This friction induces shear stress in the rock, which eventually exceeds the inherent rock strength, resulting in slip that releases elastic strain energy in the form of shock waves. These seismic waves propagate throughout the earth, and when they reach the earth's surface move the ground vertically and horizontally. This ground movement causes accelerations that result in the inertial response of structures, which in turn, causes significant damage or collapse of countless structures. There are also segments of the plate boundaries that experience smooth continuous relative deformation, or creep, such as the aforementioned segment of the San Andreas Fault along the Cienega Valley near Hollister, California (Figure 1.1a).

Seismic waves that radiate from the location where a fault ruptures (the *focus*) quickly travel throughout the earth's crust, producing ground shaking when they reach the ground surface. The intensity and duration of shaking experienced at a particular site during an earthquake are primarily because of three factors, all of which will be discussed in more detail throughout this book:

1. *Earthquake size* (magnitude): It can be measured objectively or subjectively—larger earthquakes cause stronger shaking. A strong earthquake can cause ground shaking over widespread areas, suddenly affecting large numbers of structures. Even relatively small earthquakes can have a significant impact on large numbers of buildings. The 2014 South Napa earthquake (magnitude 6), as shown in Figure 1.7, caused extensive damage to buildings in the town of Napa, as well as to the many wineries in the region, with a total estimated loss of approximately one billion US dollars. Fortuitously, the quake struck early in the morning when businesses were closed and sidewalks were empty; only one fatality was attributed to the quake.

2. *Location* (distance from the focus or epicenter): Generally, the closer to the epicenter, the stronger the shaking. Structures near the epicenter of a strong earthquake can experience extensive damage, in some cases partial or total collapse; see Figure 1.7. Many other examples can be found online.

3. *The subsurface materials beneath the structure*: Soft soils amplify the shaking, while rocks do not. This is the most insidious of the three factors because the site can be located at a long distance from the *epicenter* and still experience extensive ground shaking due to local soil conditions. Seismic waves travel through rock for most of their trip from the focus to the surface; however, at many sites, the final part of the trip is through soil, the geological characteristics of which have a major influence on the nature of ground shaking. Some soils act as seismic wave filters, attenuating shaking at some frequencies while amplifying it at others. The most peculiar example of distant seismic wave amplification is that of the Valley of Mexico, where Mexico City is located. The origin of these waves (earthquake focus) is approximately 350–400 km away, along the Pacific cost in Michoacán. Parts of this valley are situated in a dry lake, Lake Texcoco, which was emptied to make way for

FIGURE 1.7   Intensity and extent of 2014 South Napa Earthquake. (Courtesy of USGS.)

the expansion of Mexico City. The near-surface geology of the city center consists of an alluvial deposit, a thick layer of a very soft, high-water-content mixture of clay and sand that was deposited there by streams before the lake was drained (so that the Aztecs could build the island city of Tenochtitlan). As shown in Figure 1.8, rock shakes at the same frequency as the seismic waves, which attenuate as they move further from the epicenter. However, the unconsolidated alluvial soil has a dominant influence, and magnifies the intensity of ground shaking (10-fold) near certain periods, resulting in near harmonic motion with a period of 2 s. During the 1985 Mexico City earthquake (magnitude 8), peak horizontal accelerations were first attenuated from 150 cm/s^2 along the Pacific coast near the epicenter to 35 cm/s^2 at higher elevations with firm subsurface materials; these waves were then amplified fivefold to 170 cm/s^2 near the center of Mexico City. The shaking intensity was further amplified by resonant vibration properties of certain buildings, those with natural periods of about 2 s, which correspond to buildings between 10 and 20 stories, most of which experienced large displacements and extensive structural damage, or collapse.

1.2   TYPES OF EARTHQUAKES

The majority of damaging earthquakes are produced by surface fault ruptures caused by the relative movement of the tectonic plates. However, a number of other sources, including man-made and volcanic, can produce measurable earthquakes. Volcanic earthquakes can be attributed to the same forces that are responsible for tectonic earthquakes. In this section, we present a brief description of these three types of earthquakes. In Chapter 2, a more detailed description of tectonic earthquakes is provided.

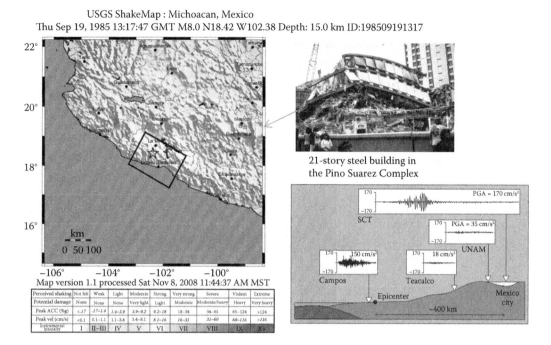

FIGURE 1.8   Intensity of 1985 Mexico City Earthquake. (Courtesy of USGS.)

1.2.1 MAN-MADE EARTHQUAKES

These generally have a much smaller magnitude than the other two types of earthquakes, and thus have a lesser impact on infrastructure. However, man-made earthquakes can lead to earlier fault ruptures (tectonic earthquakes) because the shaking can increase critical stresses at the plate boundaries. One of the most intense cases is due to explosions, both from conventional and nuclear weapons. For example, it is estimated that the Boxcar nuclear bomb explosion in 1968, with a yield of 1200,000 tons TNT equivalent, excited an earthquake of magnitude 5.0 that lasted for 10–12 s (Bolt, 2004). This shook buildings in nearby communities, including Las Vegas, NV (30 miles away), but no serious damage or casualties occurred.

Collapse earthquakes can be generated by landslides (those not generated by tectonic earthquakes) as well as collapse of mine/cavern roofs, or buildings. Earthquakes caused by mine roof collapse can be rather controversial given that tectonic earthquakes can cause mines to collapse. The Crandall Canyon, Utah coal mine collapse on August 6, 2007 was blamed on a tectonic earthquake by the owner, but seismologists determined that the 3.9 magnitude seismic shock was caused by the collapse of the mine roof. They reasoned that the coal extraction process being applied was the primary cause of the mine collapse because the center of the focus (center of the seismic activity) was in the collapsed mining zone, and most of the previous earthquakes recorded in that region were attributable to mining activities. No damage from the earthquake was detected outside the mine; however, six miners lost their lives in the accident. Although not necessarily man-made, earthquakes caused when underground caverns collapse (which sometimes leave a sinkhole on the surface) are very similar to those caused when mine roofs collapse, and are very common in many parts of the world (such as Florida). Building collapse earthquakes are rare, and even the largest buildings only cause rather small magnitude quakes. For example, the collapse of the North World Trade Center Tower on September 11, 2001 only caused a 2.3 magnitude earthquake; see Figure 1.9. This is smaller than a tectonic earthquake recorded just months earlier near the site of the tragedy.

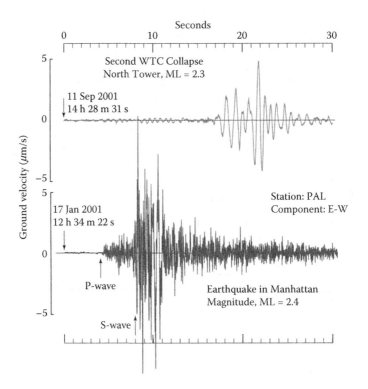

FIGURE 1.9 Comparison of building collapse and natural earthquakes. (From Kim et al. 2001. Seismic waves generated by aircraft impacts and building collapses at World Trade Center, New York City. *Eos Transactions American Geophysical Union* 82(47): 565–571.)

Seismic activity near large artificial lakes has also been reported. It is believed that the weight of the water and the extra water pressure in rocks can change the stresses in existing rock fractures. The weight of the water increases the total stress by direct loading, while the extra pore water pressure decreases the effective stress (pushes the fracture planes apart); this can lead to sliding along fracture planes where tectonic strain is already present, causing an earthquake. This is speculated to be most likely during periods when a reservoir is being filled or drained. One case that continues to be attributed to artificial lake seismicity is the August 1, 1975 Orville, California earthquake. Bolt (2004), however, presents a compelling argument against this supposition. His personal experience is that the connection between the earthquake and the filling of the lake is only circumstantial; however, he does not completely discount the possibility that the reservoir might have in part triggered the earthquake. There are other earthquakes that have been conclusively proven to have been triggered by artificial lakes; one such case is the November 14, 1981 earthquake (magnitude 5.6) induced by Lake Nasser along the Nile River in Egypt (Bolt, 2004).

Earthquakes caused by energy production processes have been recorded for over 50 years, though recently the association of earthquakes and energy production has become contentious. This is particularly controversial with respect to the hydraulic fracturing (fracking) process. This technique for extracting oil and gas from low permeability rocks involves fracturing rocks by injecting high-pressure fluid into the rock. The most compelling evidence of fracking-induced seismicity can be found in Oklahoma and north Texas, regions where seismic activities were once nonexistent, but now routinely experience small earthquakes; with the only variable being oil and gas production. In fact, a 2012 report by the U.S. National Research Council concluded that some (not many) production activities have induced perceptible seismicity. Geothermal power generation has also been conclusively correlated with earthquakes. The process entails injecting water at high pressures to fracture 500°F solid rock, creating an artificial reservoir of superheated water, the steam of which is then used to drive electrical turbines. A well-documented case of induced seismicity is from Geysers Geothermal field in California.

1.2.2 VOLCANIC EARTHQUAKES

These are caused by the same energy source as tectonic earthquakes, which is the heat from the earth's core. Volcanic seismicity affects limited areas near volcanic regions. For example, Hawaii's big island has a very active volcano that has caused several earthquakes, none of which have been felt by the entire island. These occasionally serve as a warning sign of impending volcanic eruptions. The movement of magma through tubes below the volcanic vents creates pressure changes in the surrounding rock that can rupture, releasing elastic strain energy as seismic waves. These seismic waves have been successfully used to predict eruptions of volcanos such as Mount St. Helen in 1980 and Pinatubo in 1991. Other seismic waves can be induced by sudden, irregular movement of magma whose path has been obstructed, or by steady magma movement deep in the mantle. Damage from all these earthquakes is relatively minor compared with that produced by tectonic earthquakes.

1.2.3 TECTONIC EARTHQUAKES

As discussed previously, these are caused by a sudden dislocation of segments of the earth's crust, the structure of which is composed of plates (large and small) known as *tectonic plates* that float on a liquid layer, the mantle. This arrangement resulted from the formation of planet Earth five billion years ago, when hot gasses cooled into a semi-solid mass. It is estimated that after one to two billion years of cooling, the crust solidified and cracked forming tectonic plates (different ones than those that exist today).

From the beginning, the plates have been in constant motion forming and breaking up continents over time, including the formation of supercontinents that contained most of the landmass. The latest supercontinent, Pangea, started separating approximately 200 million years ago, and its parts have drifted apart to the current configuration of the earth's surface. This process was originally proposed by Alfred Wegener in the early 1900s. He noted several different pieces of evidence to support his theory of the continental drift, including (see Figure 1.10):

1. How the current shape of some continents appear to fit together, particularly the east coast of South America and the west coast of Africa.
2. The significant similarities between fossil records (both flora and fauna) found in several continents that could only have occurred if the continents were attached.
3. The similarity in geology across several continents, including grooves carved by glaciers and the sediments deposited by these glaciers.

Wagner's theory implies that the earth's crust (known as the lithosphere) consists of large slabs of rigid rock, or plates (see Figure 1.11). The lithosphere can be divided into oceanic and continental parts. The oceanic lithosphere is thinner and forms near mid-oceanic ridges where plates are drifting apart. The thicker continental lithosphere plates consist of rocks that are less dense than oceanic lithosphere rocks. All plates move at different speeds and different directions, spreading apart along some of the plate boundaries (spreading or divergent zones), while colliding along other boundaries (subduction or convergent zones). The spreading zones are continuously replaced with upwelling lava, creating new seafloor or oceanic lithosphere. At the opposite side of the newly created oceanic lithosphere, the old part of the plate plunges into the earth's interior at the subduction zones, being

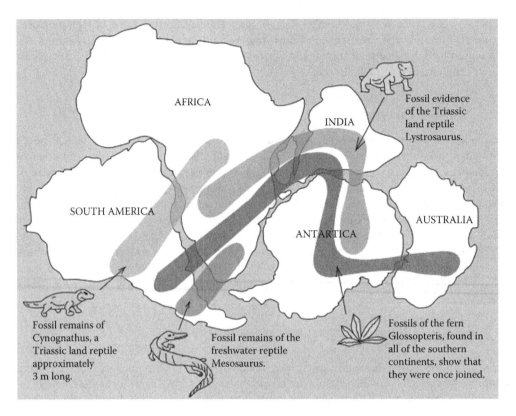

FIGURE 1.10 Continental drift evidence; distribution of fossils across the southern continents of Pangea. (Courtesy of USGS.)

absorbed into the mantle. At some locations along the mid-oceanic ridge, horizontal slip occurs normal to the divergent zones as a new surface is created or transformed; this type of boundary is known as a transformed fault (see Figure 1.11).

Wagner originally proposed that the forces driving the continental drift were the result of gravity, and the same forces were responsible for ocean tides. This theory was rejected because these forces are too small to drive the continents through the seafloor. It was later proposed that seafloor spreading at the mid-oceanic ridges was responsible for moving the seafloor and continents over the mantle, which is a softer rock immediately below. Although this theory was also rejected, it helped establish the theory of plate tectonics, which with respect to the shape and size of the plates is widely accepted today; see Figure 1.11. For a while, it was believed that the movement of the plates was driven by convection currents in the mantle; this transfers heat from deep within the earth (the core) to the surface. The temperature difference drives molten lava currents (similar to how heat from the sun drives wind currents on the earth's surface). The currents are produced as heated rock near the core (which has a lower density than that near the bottom of the plates) rises toward the surface as molted lava; over the course of the trip, lava cools and becomes denser, some of which after reaching the surface plunges back toward the core.

This theory, however, does not fully explain all phenomena associated with lithospheric plate movement. The current view is that plate movement is driven by additional forces, known as ridge push and slab pull, which also derive their energy from the dissipation of heat from the earth's core through convection currents (this theory is still being debated). These forces are generated at the mid-oceanic ridge (divergent zones) and at the subduction zones. At the divergent zones, lava cools at the surface creating new oceanic ridge; as this new material moves away from the divergent

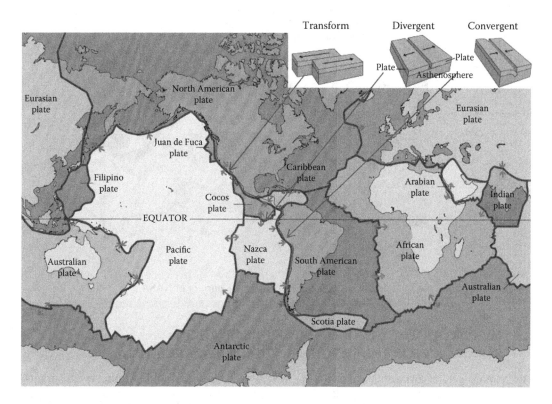

FIGURE 1.11 Tectonic plates and types of plate boundaries. (Courtesy of USGS.)

zones, it continues to cool, becoming denser. At the other end of the plate, the oceanic lithosphere becomes even denser with age as it continues to cool and thicken. This causes the lithospheric plate to tilt toward the subduction zones, like an unbalanced boat in water. The mid-oceanic ridge then rises above the rest of the seafloor, and as the new material rises to the top of the ridge, gravity pulls it down; this is called ridge push. Also, the greater density of old oceanic lithosphere at the subduction zones, relative to the underlying asthenosphere, allows gravity to pull these rocks (known as slabs) down into the mantle. This is believed to be the source of the driving force for the majority of plate movement.

Although the precise force mechanisms causing movement of the lithospheric plates continue to be debated, there is substantial agreement that heat from the earth's core serves as the primary source of energy for these forces. Also, the general outlines of the major plates have almost conclusively been established. In fact, this outline is clearly illustrated when the location of the epicenters of recent earthquakes is plotted on a global map as shown in Figure 1.12. At the boundaries of the plates, rocks fracture, usually at many locations, creating a web of smaller plates with edges that rub and push relative to each other; these edges are called *faults*. On average, these faults have the potential to displace approximately 2 inches per year. When the rubbing and pushing is prevented, elastic energy accumulates along the edges of the plates. When this energy is released with a sudden movement (*slip*), it causes brief strong ground vibrations. The specific location (generally a volume of rock) where the movement or energy release occurs is known as the *focus,* or *hypocenter.* The point on the earth's surface directly above the *hypocenter* is called the *epicenter.* Usually, the vibrations cause the rocks near the focus to become unstable; and as these rocks settle into a new equilibrium state they cause *aftershocks.* The discipline that studies seismic activity is known as *seismology,* and will be discussed further in Chapter 2.

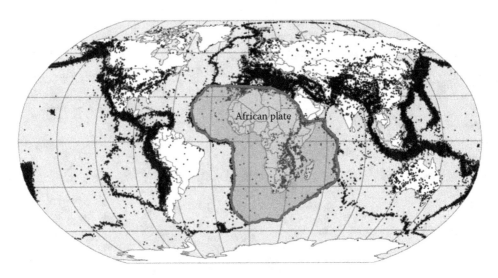

FIGURE 1.12   Epicenters of 358,214 earthquakes, 1963–1998. (Courtesy of NASA.)

1.3   HISTORY OF THE DEVELOPMENT OF MITIGATION STRATEGIES FOR THE EFFECTS OF SEISMIC HAZARDS

The primary goal of an earthquake engineer is to mitigate hazards resulting from seismic events. As previously discussed, it is difficult to mitigate the effects of tsunamis, landslides, and fires; however, mitigating the effects of ground shaking on structural systems using sound engineering judgment has made great advances in the last century. This is known as earthquake-resistant design and has been successfully applied to reduce human and economic losses. The remainder of the book focuses on earthquake-resistant design for structures. In this section, we briefly discuss the historical development of earthquake-resistant design, which chronologically has occurred in five general stages:

1. Development of construction practices to mitigate the effects of shaking.
2. The first half of the twentieth century focused on characterizing the effect of shaking using lateral forces.
3. In the 1940s, the development of the response spectrum theory marked a major step forward.
4. In the 1960s, concepts of structural dynamics were incorporated into design practice.
5. The last stage has implicitly been included in building codes since the 1970s and deals with performance-based design.

While humans did not fully understand the process that causes earthquakes until the middle of the last century, many civilizations developed construction practices to reduce the structural damage caused by ground shaking. From the study of prehistoric buildings in seismic-prone regions, it appears that two approaches to seismic-resistant construction were followed: increase the strength or change the stiffness of the lateral force-resisting systems. This is particularly evident for the construction of important structures, which received special seismic-resistance detailing, while most other buildings did not.

In many parts of the world, buildings were built stronger after they collapsed in an earthquake, while other parts of the world addressed weaknesses in the construction by specifying different construction detailing. For example, the Incas in Peru were master builders that created ingenious earthquake-resistant structures. The hilltop city of Machu Picchu (classified as a Civil Engineering

FIGURE 1.13   Seismic-resistant construction in Machu Picchu, Peru.

International Historic Monument by the American Society of Civil Engineers [ASCE]) has numerous structures constructed of interlocking, mortar-free stonework with perfectly mated joints (these are known as dry-stone walls or ashlar masonry); see Figure 1.13. Clearly, this type of construction was deliberate in order to prevent sliding of the stone blocks during earthquakes. The intricate process was far more complex than the perfectly regular array of stonework used in other parts of the world, indicating that the extra effort was purposefully made for structural reasons and not for aesthetics, as some have suggested. In fact, there are buildings in Machu Picchu with masonry laid in parallel courses (see Figure 1.13); the most important structures, such as temples and the king's quarters, were built to withstand large earthquakes using masonry laid in nonparallel courses (see Figure 1.13).

Ashlar masonry is more seismic resistant than using mortar because higher friction forces develop between mating surfaces. This along with the interlocking masonry allows the stones to move slightly and resettle without excessive movement of the walls. This type of construction allows for passive energy dissipation, a form of passive structural control. Inca construction was based on designs that follow prescriptive specifications. This type of design philosophy is based on experience with the construction of proven designs, and is still widely used for design today. More rational construction procedures based on engineering principles (engineered construction) have only been available for the past century.

Engineered seismic-resistant design was developed in America, Asia, and Europe; not necessarily independently, but there is no consensus on the order. It is believed that the first rational attempts to understand the effects of earthquakes on structures were made in Europe following the devastating 1755 Lisbon earthquake (over 50,000 casualties). After this event, the regional government enacted building regulations that required seismic-resistant detailing to be followed in the reconstruction of the city. A few years later, scientists thoroughly documented the effects of the 1783 earthquake in the Calabria region of southern Italy (nearly 50,000 casualties). Although the results from these studies did not lead to substantive design guidelines, they did inspire future scientists to study the effects of earthquakes on structures, including a number of British and Italian engineers who went on to develop the foundation for seismology.

After the 1881 Nobi earthquake (7000 casualties), the Japanese government formed the first-ever earthquake investigation committee, and directed the committee members to research earthquake-resistant structural design. It is believed that this committee conceived the idea of characterizing the effects of earthquakes on structures by applying a lateral force equal to a fraction of the total weight of the building, which is based on Newton's second law (force equals mass times acceleration). Following the 1908 Messina-Reggio earthquake (58,000 casualties), the Italian government also appointed a committee to develop practical recommendations for the seismic-resistant design of buildings. This committee also proposed characterizing the effects of earthquakes on structures by applying lateral forces: first, they proposed a force equal to 1/12 of the building weight. Three years later, this specification was modified as follows: the first story must resist a force equal to 1/12 of the building weight above, while the second and third floors must resist a lateral force of 1/8 of the building weight above. This specification limited the height of buildings to three stories. These two committees provided a first iteration to the equivalent static procedure that is widely used in seismic-resistant design codes around the world today.

In America, the first attempt to characterize seismic effects on buildings also followed a major seismic event, the 1906 San Francisco earthquake. Resistance to earthquake forces was prescribed in code provisions indirectly by requiring buildings to resist a wind lateral load of 30 lb/ft^2, which was intended to account for both wind and seismic effects. Following another devastating earthquake in California (Long Beach in 1933), the Division of Architecture was given the responsibility for public-school building safety. Their proposed regulations required that public-school buildings be designed to resist a lateral force equal to a fraction of the dead load and a portion of the design live load, entailing 0.1 for masonry buildings without frames and 0.02–0.05 for other types of buildings. This appears to be the first explicit design specification incorporating the importance of construction regarding the type of lateral force-resisting system. This also marked a very important point in code enforcement; the California Legislature passed the Field Act and the Riley Act. The Field Act made the seismic-resistant design and construction of public-school buildings mandatory, and the Riley Act required that most buildings in the state be designed to resist a lateral load equal to 2% of the design loads dead plus live loads.

The 1943 City of Los Angeles building code is believed to be the first to recognize that lateral forces depend on building mass and height, an indirect recognition of building flexibility on these forces. The lateral forces at different stories of a building were determined by the product of a gravity load (equal to dead load plus half of the live load) and a lateral force coefficient that varied with the number of stories above the story being analyzed. The 1947 San Francisco building code also recognized the influence of building flexibility on the magnitude of seismic forces, as well as the influence of site soil conditions on these forces.

Although many city building codes have been in existence around the world for decades (even millennia, such as the Hammurabi Code from Babylon, which did not include specifications, rather promoted good construction by prescribing penalties for failures), regional model codes were not developed until the introduction of the Uniform Building Code (UBC) in 1927. The second edition of the UBC published in 1930 incorporated seismic provisions and suggested these be incorporated by cities in areas prone to earthquakes. The UBC seismic provisions were based primarily on two observations of building performance during earthquakes: first, most damage was caused by lateral shaking, and second, structural damage was more extensive for buildings on soft soil deposits (likely the first code to recognize soil effects). The code specified a lateral force equal to 10% of the dead plus design live loads, reducible to 3% for buildings on firm soil. In the ensuing editions of the UBC, maps of the United States depicting zones of different levels of seismic risk were published. These maps recognized the hazard of large magnitude earthquakes for different areas of the country.

In 1948, a joint committee of the Structural Engineers Association of Northern California and the San Francisco Section of ASCE proposed model lateral-force provisions for California building codes. A major advance was the introduction of the earthquake-response spectrum concept, which was first developed by Biot in 1943 and later proposed as a design tool by Housner. This theory combines ground motion and the dynamic properties of the structure (namely the period). This led to the building period becoming an explicit factor in the determination of seismic design forces. The response spectrum work was made possible by recordings of strong-motion earthquake accelerations using recently deployed accelerographs by the US Coast and Geodetic Survey. In 1952, a second joint committee of ASCE and now the Structural Engineers Association of California (SEAOC) developed a comprehensive guide, the Recommended Lateral Force Requirements and Commentary (the so-called SEAOC Blue Book), first published in 1959. This explicitly related lateral forces to building periods based on the design response spectrum.

Structural dynamic concepts were further developed in the 1960s and incorporated in seismic codes after the 1971 San Fernando earthquake. In particular, structural dynamic specifications explicitly incorporating the response spectrum theory were required for the design of hospitals and other critical facilities. Since the incorporation of structural dynamics in code specifications, significant advances have been made in the development of analytical procedures; much of this has been in parallel with the development of computers because solutions based on structural dynamics are computationally intensive. Even with the great advances in the development of mitigation strategies for ground shaking, structures designed using modern seismic-resistant design principles have been observed to experience much larger forces during major earthquakes than those prescribed by design codes, leading to major damage, even collapse. These observations have led researchers to conclude that structures subjected to larger imposed loadings from earthquakes can survive intense shaking with damage; however, will not collapse as long as lateral force-resisting systems are tough (i.e., able to absorb the seismic energy) and include a continuous path for lateral load to be transferred to the ground. Also, it is important for engineers to understand that the structural engineering principles used for the analysis of structures subjected to gravity static loads do not result in the greatest dynamic forces a structure may experience during an earthquake. These dynamic forces are only equivalent to code-specified lateral forces in that a structure designed to resist these forces has the capability of deforming without overstressing from load reversals, and provide adequate member ductility, as well as provide connections with sufficient strength and resiliency to accomplish the following performance goals:

- Resist minor earthquakes without damage.
- Resist moderate earthquakes without structural damage, but with some nonstructural damage.
- Resist major earthquakes without collapse, but with both structural and nonstructural damage.

These performance criteria have generally remained unchanged since the 1970s, but have recently taken on a broader approach in order to allow the real estate owner to decide the desired level of structural performance and safety; both in terms of protecting occupants and capital investment. This new philosophy is known as performance-based design and is awaiting to appear in standards such as the 2010 ANSI/ASCE "Minimum Design Loads for Buildings and Other Structures" standard; see Section 1.3.1.3 in this document.

REFERENCES

Bolt, B. A., *Earthquakes*, 5th edition, W.H. Freeman and Co., New York, 2004.
Wald, D., Quitoriano, V., Heaton, T., and Kanamori, H., Relationships between peak ground acceleration, peak ground velocity and modified mercalli intensity in California, *Earthquake Spectra*, 15(3), 557–564, 1999.

2 Engineering Seismology Overview

After reading this chapter, you will be able to:

1. Define seismology terminology
2. Describe tectonic plate movement based on the elastic rebound theory
3. Describe different types of seismic waves
4. Describe how earthquakes are measured based on intensity and magnitude
5. Describe major effects of earthquakes on structures
6. Describe earthquake characteristics
7. Describe basic site characteristics
8. Describe structural characteristics
9. Compute the probability of occurrence of an earthquake

Seismology is the scientific discipline that studies seismic activity from various sources (tectonic, volcanic, etc.), in particular, the propagation of seismic waves through the interior of the earth and their environmental effects, such as ground shaking and tsunamis. This discipline grew out of the need to understand how earthquakes occur, and to try and predict their occurrence. Engineering seismology is the applied branch of seismology that focuses on characterizing seismic hazards at a site or region in order to evaluate risks to various vulnerable infrastructure systems. This is accomplished by studying the seismic history of a region (including the frequency and intensity of earthquakes) to establish characteristics (such as the intensity of shaking) and frequency of occurrence of future earthquakes. In this chapter, we briefly cover the most important concepts of engineering seismology.

2.1 SEISMOLOGY TERMINOLOGY

Just like any other discipline, seismology has developed a jargon as a means to communicate succinctly. In this section, we introduce and define terminology related to earthquake engineering. One of the most recognized terms related to seismology is *epicenter*, which is the geographical point on the ground surface where an earthquake is estimated to be centered. This point along with the *focal depth* gives the location where the rock ruptures at a fault (fault rupture) that generates the main earthquake, the *focus* or *hypocenter*. This is an area (not a point) that can extend for many miles along a fault. The plane along which the rock ruptures and slips is known as the *fault plane*, and may be at an angle with respect to the ground surface, the *dip angle*. The angle the fault plane makes with respect to the north direction along the surface is known as the *strike angle*. The relative displacement between the two sides of the fault plane is known as the *fault slip*. The radiating seismic waves (discussed in Section 2.1.2) are recorded using a seismometer at an *observation station* located at a distance known as the *epicentral distance*. These terms are used to characterize earthquakes and are illustrated in Figure 2.1.

The *focal depth* is used to classify earthquakes as: *shallow* for focal depths less than 70 km (43 miles), *intermediate* for focal depths between 70 km (43 miles) and 300 km (186 miles), and *deep* for focal depths greater than 300 km (186 miles). Focal depths have been known to be as large as 720 km (450 miles). Shallow earthquakes are more destructive than others because the mass of the rock above deeper earthquakes attenuates their shock waves. Shallow earthquakes generally occur near ocean trenches, like coastal California earthquakes, which have focal depths of less than

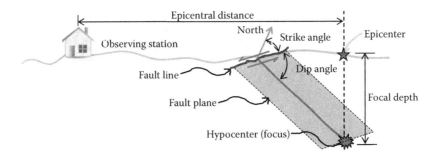

FIGURE 2.1 Earthquake terminology.

16 km (10 miles). Most moderate-to-large shallow earthquakes are preceded by smaller quakes, called *foreshocks*, and followed by smaller quakes called *aftershocks*. Foreshocks are precursors of the impending fault rupture, while aftershocks result from adjustments to the stress imbalance in the rocks produced during the rupture.

The *fault slip* is used to classify faults depending on the direction of the movement of rocks on one side relative to the other side; vertical movement is known as *dip-slip*, while horizontal movement is known as *strike-slip*. These terms correspond to the definition of the angles used to describe the direction of fault movement. As discussed in Section 1.2.3, faults occur along the edges of tectonic plates (or their interior, which is much less common). Dip-slip faults are further subdivided based on the relative vertical movement of the tectonic plates: *normal* faults occur at plate boundaries where tectonic plates spread apart at divergent zones causing the *hanging wall* to move down relative to the *footwall*, while *reverse* faults occur at plate boundaries where tectonic plates collide at convergent zones causing the *hanging wall* to move up relative to the *footwall*; see Figure 2.2. *Thrust* faults are reverse faults with a small dip angle. Faults along oceanic ridges are normal, resulting in the lengthening of the crust, as shown previously in Figure 1.11, whereas those along subduction zones are reverse, resulting in the shortening of the crust. When these faults break through to the surface, *surface rupture*, they produce an exposed steep slope known as the fault's *scarp*; see Figure 2.2.

At transform zones, tectonic plates move horizontal relative to each other; the faults associated with this motion are strike-slip, which can be further subdivided into *right-lateral* or *left-lateral* faults. Figure 2.3 depicts a left-lateral fault, which can be described by an observer standing in front of the fault line as watching the land on the other side moving to the left. Similarly, an observer facing a right-lateral fault line would see the land on the other side moving to the right. The San Andres fault discussed in Chapter 1 is a type of right-lateral fault. This is demonstrated by the fault extending from left to right through the drainage culvert on the south side of the DeRose Winery building in the Cienega Valley; see Figure 2.3. Therefore, when these faults break through to the surface, any pipes, streams, fences, etc. that straddle the *surface rupture* are offset. It should be noted that dip-slip and strike-slip faults can occur simultaneously, in which case the fault is classified as an *oblique-slip*. Also, fault rupture does not reach the surface in the case of many earthquakes, particularly for intermediate and deep cases.

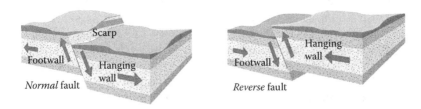

FIGURE 2.2 Vertical movement, *dip-slip* faults.

Right-lateral fault

Left-lateral fault

FIGURE 2.3 Horizontal movement, *strike-slip* fault.

2.1.1 ELASTIC REBOUND THEORY: TECTONIC PLATE MOVEMENT

As discussed in Section 1.2.3, tectonic plates are forced to move by several mechanisms, including ridge push and slab pull, convection currents in the semimolten rock of the mantle, etc. At the boundaries, relative movement between plates tends to occur; however, shear stresses develop along the plane that separates the fault because of friction between the plate boundary surfaces. As these shear stresses reach the shear strength of the rock along the fault, the rock fails and elastic strain built-up in the rock is released as the rock fractures and slips. The rate at which the stored strain energy is released is dependent on the geometry and material properties along the fault, and it dictates whether the resulting movement is aseismic (in the form of creep; see Figure 1.1) or seismic (in the form of an earthquake). Along relatively straight stretches of fault where the rock is weak and ductile, the rock tends to rupture slowly releasing the strain energy gradually, which results in creep. Where the rock is strong and brittle, it ruptures fast, releasing the strain energy abruptly—partly in the form of heat and partly as radiating stress waves that cause earthquakes on the ground surface. For this case, after the strain energy is released, there is a period of no seismic activity until the subsequent buildup of strain energy occurs followed by rupture. Therefore, earthquakes are more likely to occur along parts of a fault line that have not ruptured for some time; these are known as *seismic gaps*, which can be identified by plotting seismic and creep activity along a fault line (Kramer, 1996). This process of successive buildup and release of strain energy is described by *elastic rebound theory* and can be described graphically, as shown in Figure 2.4. The figure depicts the gradual buildup of strain resulting from the relative motion of two plate boundaries when the motion is being resisted by friction forces. The fault slips when the stresses overcome the shear strength of the rock, and the two sides of the fault rebound to an unstressed state, resulting in radiating seismic waves.

2.1.2 SEISMIC WAVES

Seismic waves radiate from the focus and travel in every direction, as shown in Figure 2.5. The portion of the energy released from a fault rupture as shaking first travels through the interior of the earth as *body waves*, following the shortest path to the ground surface where they are transformed into *surface waves* (see Figure 2.5). Thus, body waves arrive at an observation station (or site) before any surface waves. An understanding of these different waves is necessary to establish the epicenter and to characterize the size of the earthquake (earthquake magnitude and ground acceleration).

Unstrained rock   Strain buildup   Rebound to unstrained state

FIGURE 2.4 Elastic rebound theory.

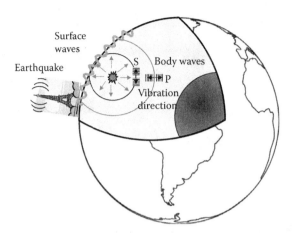

FIGURE 2.5   Seismic waves.

Body waves naturally divide into two different types, P-waves (*primary* or *pressure*) and S-waves (*secondary* or *shear*), which are slower than the P-waves; see Figure 2.5. The stiffness of the material through which the waves travel controls their speeds. P-waves are longitudinal waves that move through earth's interior in successive compression and rarefaction (i.e., P-waves move material particles back and forth parallel to their direction). Their average speed through solids is approximately 6 km/s and slows down when traveling through liquid, for example, speeds in water are only about 2 km/s. Also, given that liquid medium (such as water) has no shear stiffness, S-waves can only travel through solids, at about 3 km/s. Because of the slower speeds of S-waves, they travel at a lower frequency, but larger amplitude, which makes them more destructive than P-waves. This speed differential between P- and S-waves can be recorded using a seismograph as a time lapse, Δt_{p-s} (see Figure 2.6), and allows the seismologist to predict an impending earthquake at a site, albeit with a very short warning. This was the basis for an earthquake early warning system implemented in Mexico City a few years after the 1985 earthquake. Because of the large distances (approximately 400 km; see Figure 1.8) from the epicenter to the city, it is estimated that the residents of the city have 60–90 s (the time it takes between the arrival of the P-waves and the intense shaking from S-waves) to prepare for the impending earthquake.

When body waves reach the ground surface, they are classified as surface waves. These have lower frequency than their parent body waves and are obviously responsible for the damage associated with earthquakes. The intensity of these waves diminishes approximately exponentially as the focal depth increases. These waves are also divided into two types: Rayleigh and Love. Rayleigh waves combine effects of both P- and S-waves on the earth's surface and move disturbed material vertically and horizontally (in the direction of wave propagation) simultaneously, similar to ocean waves. These waves have low frequencies and large amplitudes, and travel at average speeds of

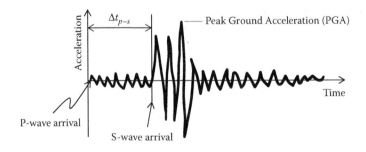

FIGURE 2.6   Seismograph—a strong-motion recording using a seismometer.

3 km/s. Love waves move the ground side to side on the surface plane, perpendicular to the direction of the wave propagation, similar in character and speed to S-waves. The shear motion of these waves is primarily responsible for the ground shaking experienced during earthquakes.

As shown in Figure 2.6, seismic waves can be recorded using a seismograph, which can be used to decipher the four different types of waves since each travels at different speeds, with P-waves being the fastest, followed by S-waves, then Love waves, and finally Rayleigh waves. Along with the direction of shaking on the ground surface (vertical and horizontal), the different speeds relative to one another can be used to separate these waves from one another in a predictable pattern; this is particularly clear for P- and S-waves in Figure 2.6. For surface waves, the vertical and horizontal (north–south and east–west) recordings are needed to decipher the two surface waves. The different arrival times of these waves can be used to estimate the epicentral distance (and the hypocentral distance—the distance from the observation station to the focus). With recordings from three different observation stations, an estimate of the geographic location of the epicenter can be triangulated. That is, the epicenter is located where three circles drawn on a map intersect, each circle of radius equal to the epicentral distance, and centered at their corresponding observation station location.

2.2 MEASURING EARTHQUAKES

The duration and size of an earthquake can be measured using several objective and subjective methods, including relative strength (shaking intensity), amount of energy radiated during the fault rupture, and the magnitude of the largest acceleration imparted to a site or a structure. The shaking intensity measures earthquake size using a single quantitative measure, either the intensity (subjective) or magnitude (objective); see Figure 2.7. These two will be discussed further in the next two sections and are typically used to compare events of different sizes quantitatively. Energy released is measured objectively using one of a number of methods to be discussed later. The acceleration is quantified at a site using the peak ground acceleration (PGA), which is defined as the maximum ground motion (see Figure 2.6).

2.2.1 EARTHQUAKE INTENSITY

The intensity of an earthquake is a subjective, nonempirical approach for estimating the size of an earthquake based on the subjective assessment of human observations of the effects of earthquake shaking on buildings (amount of damage sustained by structures and land surface) and on people. Different buildings (based on type and construction material) respond differently to shaking; that is, the performance of poorly designed and constructed buildings is more sensitive to shaking than

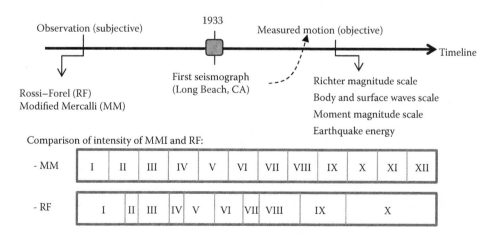

FIGURE 2.7   Scales used to estimate the relative size of earthquakes.

well-designed and/or well-constructed structures. Also, people have varying perceptions of shaking, that is, psychologically some people are more sensitive to shaking than others. This subjective measure of earthquake intensity was developed before any instruments were capable of recording earthquake shaking, and thus its value has diminished since the development of instrumental earthquake size measurements. However, the intensity methodology is still useful for assessing historic events from descriptions of damage to buildings and land, particularly given criteria that account for construction quality, geologic conditions, and potential discrepancies between observers of the same earthquake.

One of the earliest criteria (or scales) was a 10-degree scale proposed in 1883 by two scientists for whom the scale was named the Rossi–Forel (RF) scale. This is an arbitrary scale that characterizes shaking numerically in Roman numerals from 1 (RF-I) for barely noticeable to 10 (RF-X) for an extremely intense earthquake; see Figure 2.7. This scale was not widely accepted; however, it is still occasionally used in some parts of the world. In 1902, Giuseppe Mercalli proposed a modified version of the RF scale that found wide consensus amongst seismologists. The main issue with the RF scale was that the 10th degree was too broad (see Figure 2.7). The new scale still had 10 degrees, but intensities were distributed more evenly, with more detailed and explicit definitions of each degree. This new scale was later modified by a number of seismologists, including Richter who modified it to better account for contemporary construction in California, and was published in 1931 as the modified Mercalli (MM) scale (Richter, 1958). This is still widely used today, and it has 12 points, ranging from 1 (MM-I) for imperceptible shaking to 12 (MM-XII) for total destruction; see Table 2.1. It is important to note that even this scale has many limitations because it is based

TABLE 2.1
Modified Mercalli Intensity Scale, and Comparison to Magnitude and PGA

Degree	Intensity Description	Magnitude	PGA (g)
I	Not felt, except by some in rare circumstances.	<3.0	<0.03
II	Felt only by a few persons at rest on upper floors of buildings.	3.0–3.9	
III	Felt indoors, but many people do not recognize it as an earthquake.		
IV	During the day felt indoors by many, outdoors by few, while at night some awakened. Dishes, windows, doors are disturbed and walls make creaking sounds.	4.0–4.9	
V	Felt by nearly everyone during the day and many awakened at night. Some dishes, windows, etc. are broken; cracked plaster in a few places; unstable objects overturned.		0.03–0.08
VI	Felt by all, many frightened and run outdoors. Some heavy furniture moved; a few instances of fallen plaster and damaged chimneys. Damage slight.	5.0–5.9	0.08–0.15
VII	Everybody runs outdoors. Damage negligible in buildings of good design and construction; slight to moderate in well-built ordinary structures; considerable in poorly built or badly designed structures; some chimneys damaged.		0.15–0.25
VIII	Damage slight in specially designed structures; considerable in ordinary buildings (some with partial collapse); great in poorly built structures. Chimneys, factory stack, columns, monuments, walls toppled. Heavy furniture overturned.	6.0–6.9	0.25–0.45
IX	Damage considerable in specially designed structures; well-designed frame structures thrown out of plumb; great damage for others, including partial collapse. Buildings shifted off foundations. Ground cracked conspicuously. Underground pipes broken.	7.0–7.9	0.45–0.6
X	Some well-built wooden structures destroyed; most masonry and frame structures destroyed. Ground badly cracked.		0.6–0.8
XI	Few masonry structures remain standing. Bridges destroyed. Broad fissures in ground. Underground pipelines completely out of service.	>8.0	0.8–0.9
XII	Damage total. Waves seen on ground surface. Objects thrown into the air.		>0.90

on a subjective assessment of observations (observer's previous experience with earthquakes), local design and construction practices, and population density. Furthermore, the MM scale considers the response of a building to ground shaking, not the ground shaking alone. Shaking intensity is highly affected by regional and near-surface geology (soil/rock type, etc.), and it is difficult to compare deep and shallow event intensities because deep events produce smaller levels of shaking for the same size earthquake. Table 2.1 also shows a comparison of approximate magnitudes and PGA ranges that would be expected for shaking near the epicenter.

Using either scale, various sites can be assigned intensity values and can also be used to develop an isoseismal map for a particular event. That is, sites with equal intensity values are connected using contour lines, similar to those that represent data on a topographic map. This type of map is useful in estimating the general location of the epicenter of an earthquake and its magnitude (as shown in Table 2.1 for MM), particularly earthquakes that predate instruments. Additional information that can be inferred from these maps include the ground conditions at a particular site, the underlying geology, radiation pattern of seismic waves, and response of different types of buildings.

2.2.2 EARTHQUAKE MAGNITUDE

Whereas the intensity of a given earthquake varies from one observation point to another, earthquakes can be associated with a single value of magnitude. The consensus measure of magnitude is based on the Richter scale, which quantifies the size of an earthquake with an index of the amount of energy released, and while this approach is an improvement as compared to the intensity scales, the Richter scale does not accurately account for all factors that contribute to the actual size of an earthquake. This scale, however, does appropriately measure the relative strength of an earthquake and remains an important parameter in earthquake hazard analysis. In addition to being used for earthquake hazard analysis by seismologists and engineers, it is the preferred scale used to inform the public of the size of an earthquake.

Because it was originally developed to quantify the strength of Southern California earthquakes, the Richter scale is also known as the local magnitude scale, M_L. Richter defined M_L using the base-10 logarithm of the peak trace amplitude (in micrometers, μm) of a standard Wood–Anderson seismograph (which has a magnification factor of 2800, a natural period of 0.8 s, and damping of 80%), located on firm ground at a distance of 100 km from the epicenter. The following relationship gives the local magnitude:

$$M_L = \log_{10} \frac{A}{A_o}$$

where:

A is the peak amplitude in μm, measured from a seismogram

A_o is the peak amplitude of a zero-magnitude earthquake in μm, which is used to adjust the variation of ground motion amplitude for epicentral distances other than 100 km

This equation is typically not used directly to determine the magnitude of an earthquake; instead, a correction nomogram provided by Richter is used (see Bolt 2004). The nomogram was convenient when digital calculators were not available. The equation used to develop the nomogram is

$$M_L = \log_{10} A + 3\log_{10}(8\Delta t_{p-s}) - 2.92 \tag{2.1}$$

where:

A is now given in millimeters (mm)

Δt_{p-s} is the time between the arrival of P- and S-waves in seconds (see Figure 2.6); this indirectly measures the epicentral distance

The Richter scale has a practical range from 0 to 9.0 (theoretically, the scale has no upper or lower limits). Also, the base-10 logarithmic scale indicates that each unit increase in M_L corresponds to a 10-fold increase of the earthquake wave amplitude. For example, a 7 M_L earthquake is 100 times stronger than a 5 M_L event ($10 \times 10 = 100$).

EXAMPLE 2.1

Estimate the local magnitude of a southern California earthquake recorded in two perpendicular directions at several stations using standard Wood–Anderson seismographs. The trace amplitudes and epicentral distances are as follows (Richter, 1958):

Station	Amplitude on the Seismograph, A (μm)		Epicentral Distance, Δ (km)
	N–S Component	E–W Component	
1	8400	6000	114
2	7900	8500	179
3	24,500	30,000	90
4	8100	7000	246

SOLUTION

Equation 2.1 can be rewritten in a slightly different form to include the epicentral distance rather than Δt_{p-s}, $M_L = \log_{10} A + 2.56 \log \Delta - 5.12$ (this type of equation is dependent on the seismological station and still yields significant scatter of the data). This equation can be used to determine M_L for the two perpendicular directions, which can then be averaged to obtain a sufficiently accurate result without having to combine the results vectorially. The calculations for Station 1 in the north–south and east–west directions are as follows:

$$M_{LN-S} = \log_{10} A_{N-S} + 2.56 \log_{10} \Delta_{N-S} - 5.12$$

$$M_{LN-S} = \log_{10}(8.4 \times 10^3 \ \mu m) + 2.56 \log_{10}(114 \ km) - 5.12 = 4.07$$

$$M_{LE-W} = \log_{10}(6 \times 10^3 \ \mu m) + 2.56 \log_{10}(114 \ km) - 5.12 = 4.92$$

The average of these two values gives M_L at Station 1. That is,

$$M_{L1} = \frac{4.07 + 3.92}{2} = 4.0$$

All other stations are treated similarly and the results are summarized as follows:

Station	Local Magnitude		
	M_{LN-S}	M_{LE-W}	Average M_L
1	4.07	3.92	4.0
2	4.54	4.58	4.6
3	4.27	4.36	4.3
4	4.91	4.84	4.9

Therefore, the magnitude of this particular earthquake can be determined as the average of the averages of each station:

$$M_L = \frac{4.0 + 4.6 + 4.3 + 4.9}{4} = 4.4$$

The local energy associated with earthquakes during fault fracture growth is primarily transformed into heat, with only 1%–10% being released as seismic waves. The relationship between the fraction of energy radiated as waves and the local magnitude, M_L, is given as follows:

$$\log_{10} E = 11.8 + 1.5M_L \tag{2.2}$$

where:

E is in ergs, which is a relatively small unit, 1 ft lb $= 1.356 \times 10^7$ ergs

In Equation 2.2, the logarithmic scale coupled with the other terms results in a 32-fold increase in energy radiated for each unit increase in M_L. For example, a 7 M_L earthquake radiates 1000 times more energy than a 5 M_L event ($32 \times 32 = 1024$). Since this relationship was calibrated using various types of seismic waves, it is also applicable to other magnitude scales, including those discussed in the following paragraphs.

The local magnitude scale does not accurately account for deep focal depths (>45 km) or long epicentral distances (>600 km) because seismic waves attenuate with distance (whether the distance is depth or length). The wave train recorded on a seismograph for a deep earthquake is very different than for a shallow one, leading to two different values of M_L even for events that release equal amounts of total energy. Consequently, more accurate measures of magnitude had to be developed in order to improve the uniform coverage of earthquake size. Richter working with his colleague Gutenberg addressed the shortcomings of the local magnitude scale by developing the *body-wave* scale, m_b, to handle deep-focus earthquakes and the *surface-wave* scale, M_s, to handle distant earthquakes. The three scales are related by the following empirical relationships:

$$M_s = 1.27(M_L - 1) - 0.016M_L^2 \tag{2.3}$$

and

$$m_b = 0.63M_s + 2.5 \tag{2.4}$$

Unfortunately, even these scales do not provide an accurate estimate for the largest magnitude ($>8M_L$), most devastating earthquakes. The limitations of the scales are purported to be the result of ground-shaking characteristics (such as the natural period) not increasing proportionally with increasing amount of the total energy radiated by very large earthquakes. M_L, m_b, and M_s are valid for natural periods of up to 0.8 s (for a Wood–Anderson seismograph), 1.0 s, and 20 s, respectively. Because of these ground-shaking limitations, earthquakes larger than certain sizes will have a constant magnitude. This is known as *magnitude saturation*. For instance, M_L begins to saturate at about 6.5, m_b at about 7.0, and M_s at about 8.0.

The strength of an earthquake can be more accurately measured using the moment magnitude scale, M_W, which accurately measures a wide range of earthquake sizes and is applicable globally. This scale is a function of the total moment release by an earthquake. The moment is a measure of the total energy released. The concept of moment is adopted from mechanics and is defined as the product of the fault displacement and the force causing the displacement. A simple derivation of the moment, M_0, is presented in Villaverde and is based on the size of the fault rupture, the slip amount, and the stiffness of the fractured rocks. That is,

$$M_0 = GA_f D_s \tag{2.5}$$

where:

M_0 is in dyne cm, which is a relatively small unit, 1 dyne cm $= 1 \times 10^{-7}$ N m

G is the shear modulus of the rocks included in the fault in dyne/cm^2, which ranges from 3.2×10^{11} dyne/cm^2 in the crust to 7.5×10^{11} dyne/cm^2 in the mantle

A_f is the area of the fault rupture in cm^2
D_s is the average fault slip or displacement in cm

All of which can be estimated relatively accurately. Alternatively, the moment can be directly estimated from the amplitudes of long-period waves at large distances, with corrections for attenuation and directional effects. Although moment is an effective way to establish the size of an earthquake, it is customarily convenient to convert it into a magnitude quantity so that it can be compared to M_L, m_b, and M_s. This can be accomplished by relating the seismic moment M_0 to the radiated energy E. That is,

$$E = \frac{\Delta\sigma}{2G} M_0 \qquad (2.6)$$

where:
$\Delta\sigma$ is the static stress drop in the earthquake, which ranges from 30 to 60 bars (1 bar = 100 kPa)
G is the shear modulus of the medium near the fault, the same as used in Equation 2.5

The moment magnitude, M_w, is then derived by substituting Equation 2.6 into Equation 2.2 and replacing M_L with M_w since the two scales give similar values for earthquakes of magnitude ranges from 3 to 5. For average values of $\Delta\sigma$ and G ($\Delta\sigma/G \cong 10^{-4}$), the relationship between M_0 and M_w is then given as

$$M_w = \frac{2}{3}\log_{10} M_0 - 10.7 \qquad (2.7)$$

where:
M_0 is in dyne cm

The resulting M_w does not saturate at large magnitude values. M_w is also known as the Kanamori wave energy, after the scientist who developed this relationship (Villaverde, 2009).

EXAMPLE 2.2

Estimate the seismic moment and moment magnitude of the January 12, 2010 Haiti earthquake. It is estimated that the *blind thrust* fault (the slip plane ends before reaching the earth's surface) caused an average strike-slip displacement of 2 m over an area equal to 30 km long by 15 km deep (Eberhard et al. 2010). Assume that the rock along the fault has an average shear rigidity of 3.2×10^{11} dyne/cm^2.

SOLUTION

1. Determine the fault's rupture area A_f and fault slip, D_s, in consistent units.

$$A_f = (30 \times 10^5 \text{ cm})(15 \times 10^5 \text{ cm}) = 4.5 \times 10^{12} \text{ cm}^2$$
$$D_s = 2 \text{ m} = 20 \text{ cm}$$

2. Determine the seismic moment M_0 using Equation 2.5.

$$M_0 = GA_f D_s$$
$$= (3.2 \times 10^{11} \text{ dyne/cm}^2)(4.5 \times 10^{12} \text{ cm}^2)(200 \text{ cm})$$
$$= 2.88 \times 10^{26} \text{ dyne cm}$$

3. Determine the moment magnitude, M_w, using Equation 2.7.

$$
\begin{aligned}
M_w &= 2/3 \log_{10} M_0 - 10.7 \\
&= 2/3 \log_{10}(2.88 \times 10^{26} \text{ dyne cm}) - 10.7 \\
&= 6.94
\end{aligned}
$$

It is worth noting that a magnitude 7 M_w has been widely reported in the media.

There are other magnitude scales in use that are not covered in this section; thus, it is important to specify the type of magnitude scale being used. In this book, M_L is used unless otherwise noted.

2.3 EFFECTS OF EARTHQUAKES ON STRUCTURES

To design structures that pose low risk to the public, engineers must understand the behavior of these systems when subjected to various loads, including seismic loads. Strong shaking during an earthquake is responsible for most damage to structures except in cases where structures are built directly on a fault line, in which case the shaking and ground rupture contribute to the damage. The severity of shaking at a site is controlled primarily by two factors: earthquake characteristics and *attenuation* of ground motion. Attenuation is the dissipation of seismic energy as seismic waves move through layers of varying soil and rock strata. Also, once at a site, seismic waves can be modified by the site characteristics. Furthermore, during an earthquake, different structures will behave differently, so characteristics of the structural system have a major influence on seismic structural performance. Therefore, damage to a structure is primarily controlled by the characteristics of the earthquake, site, and structure.

2.3.1 EARTHQUAKE CHARACTERISTICS

Although the primary factor controlling an earthquake's effect on structural behavior at a particular site is the PGA, the largest PGA earthquakes do not always cause the most damage from shaking. Shaking is rather complex and it is difficult to characterize its full effect using a single parameter. Strong shaking also significantly depends on duration and frequency content, and to a lesser extent on factors such as length of fault rupture, focal depth, orientation of the fault, speed of rupture, and whether or not fault rupture reaches the ground surface.

As discussed earlier, the size of an earthquake can be measured using various parameters (intensity, magnitude, etc.), the most important of which is PGA when it comes to seismic-resistant design. This is obtained from the acceleration time history recorded with an accelerometer (which is a recording of the variation of acceleration amplitude with time) and is typically given as a fraction of the acceleration due to gravity. PGA is generally largest near the epicenter and tends to decrease away from it because of attenuation. However, PGA can increase significantly depending on a number of factors, especially the geology of the path of the waves to a specific site and local soil conditions (to be discussed in the next section). Ground acceleration is generally resolved into a vertical and two horizontal components. The vertical component is approximately one-third of either of the horizontal components; thus, the majority of the damage is caused by the horizontal components of acceleration. Nonetheless, the vertical component of acceleration is also included in modern seismic design codes, as discussed in Chapter 8.

As mentioned earlier, large PGA earthquakes do not always create the most damage. In fact, a large PGA earthquake would produce little damage to some structures if it only occurs for a short time, whereas a relatively small PGA earthquake that continues for several seconds can be devastating to some structures. This can be exemplified using the first two earthquake records in Figure 2.8; whereas the PGA for the 1940 El Centro earthquake (PGA $= 0.32g$, $M = 7.0$) is smaller than that from the 1966 Parkfield earthquake (PGA = 0.5g, M = 6.2), the resulting damage from the El Centro earthquake was much more extensive than that associated with the Parkfield earthquake. This can largely be attributed to the much longer strong-motion duration of the El Centro earthquake. The duration of strong motion (or shaking) is important because the longer the strong motion, the more energy is imparted to the structure; and because of the limited capacity of a structure to absorb energy, there is a greater potential for inelastic response for longer strong shaking. Since inelastic damage accumulates from load reversals, an earthquake with a large number of load reversals would result in more damage compared to an earthquake with only a few larger load reversals. It is important to note that the duration of strong shaking is not the duration of the entire earthquake event, which is much longer, as can be seen in acceleration time history data in Figure 2.8. While there are a number of procedures that have been proposed to define the strong-shaking portion of an acceleration time history or accelerogram, none have garnered the consensus of the seismological or engineering communities. One such procedure for determining strong-motion duration is the

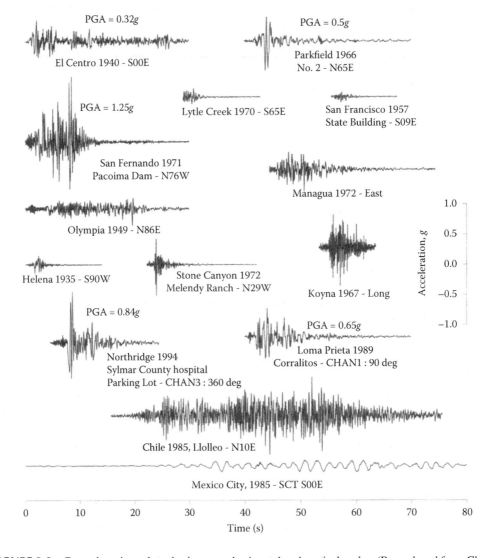

FIGURE 2.8 Ground motions plotted using same horizontal and vertical scales. (Reproduced from Chopra, A. K., *Dynamics of Structures: Theory and Applications to Earthquake Engineering*, 4th edition, Prentice-Hall, Upper Saddle River, NJ, 2012. By permission of Prentice-Hall, Upper Saddle River, NJ.)

bracketed duration, which quantifies strong-motion duration as the time elapsed between the first and last peak acceleration amplitudes that exceed a specified value (Bolt proposed 0.05g).

The dominant period at which the strong shaking occurs is often complementary and interrelated to both PGA amplitude and strong-motion duration. Small PGA amplitude earthquakes cause little damage to some structures because stresses may not be large enough, and even large PGA amplitude earthquakes with short strong-motion duration may not cause enough stress reversals to cause any significant structural damage. However, a relatively small PGA that continues for several seconds at a uniform frequency can be devastating to some structures, particularly if the dominant period coincides with the natural period of the structure, which results in a resonant condition, leading to severe damage. This is the reason for the devastating damage experienced by some buildings during the 1985 Mexico City earthquake, which had an epicenter approximately 400 km from Mexico City. The shaking in poor soil areas was devastating to 10–20-story buildings because the frequency content was filtered as the seismic waves entered these areas, leaving a dominant period of shaking of approximately 2 s (see Figure 2.8). The dominant period of shaking coincided with the natural period of 10–20-story buildings, leading to a resonant condition; interestingly, other nearby buildings suffered little or no damage.

2.3.2 SITE CHARACTERISTICS

As discussed earlier, the strength of shaking usually diminishes as waves travel away from the focus because of attenuation. Surface waves usually experience the same effect; this results in smaller shaking for sites at greater distances from the epicenter. However, shaking can also strengthen depending on the geologic conditions along the path of the wave between the focus and the site, as well as the soil conditions and topography at the site. Soil condition and topography, in particular, can greatly affect the characteristics of the input motion in terms of the amplitude and natural period of the shaking at the site. A soft subsurface deposit or steep topography (ridge or hill) causes the ground to vibrate like a flexible system, while a stiff bedrock causes the ground to vibrate as a rigid body. That is, hard rock vibrates with the same frequency and amplitude as the input motion, while the vibration of a flexible underlying deposit would depend on its inherent stiffness and damping characteristics (in the same way as the dynamic response of a structure is affected by its dynamic characteristics, as discussed in Section 2.3.3). For this case, the stiffness of the underlying material has the greatest effect on the characteristics of the input motion at a site (assuming effects of damping to be negligible). Low-frequency (long-period) input motions are amplified more at sites underlain by soft soils than those underlain by stiff ones, whereas stiff soil deposits amplify high-frequency (short-period) input motions more than soft ones.

The inherent stiffness of the ground beneath a site is a function of its *shear-wave velocity,* and thickness of the sediment above the bedrock. Shear-wave velocity is faster for hard rock than for soft soil; consequently, seismic waves travel faster through hard rock than through soft soil. So when seismic waves pass from rock into soil, they slow down. This slower speed must be accompanied by an increase in wave amplitude (amplifying the ground shaking) in order for the flow of energy to remain constant, which is required for the conservation of elastic wave energy. An increase in soft sediment thickness has a similar effect (increased amplitude) because of a decrease in material density.

The most spectacular example of a deep flexible soil deposit amplifying low-frequency motions occurred during the 1985 Mexico City earthquake. As discussed earlier (Section 1.1.3), a relatively small rock acceleration caused a high PGA at the ground surface underlain by a very soft, high-water-content mixture of clay and sand. The most severe damage was limited to an area near the city center (the so-called Lake Zone), with little or no damage to buildings built on solid volcanic rock just a few kilometers away (the so-called Foothill Zone); see Figure 1.8. The PGA for the Lake Zone was about five times greater than that in the Foothill Zone. The frequency content for the two zones was also very different; the Lake Zone had predominant periods of about 2 s, which was about 10 times larger than that in the Foothill Zone. Also, as shown in Figure 2.8, the strong shaking in the Lake Zone continued for a long time. The duration of strong shaking along with the 2 s site period contributed to a peculiar damage pattern. Minor damage occurred to buildings of less than five stories and those of more than 30 stories; most other buildings collapsed or sustained extensive damage. A simple rule of thumb for estimating the fundamental period of an N-story building is $N/10$ s (Equation 3.19), which is the reason that buildings of about 20 stories collapsed; their fundamental period matched that of the site ground motion period, creating a resonance condition. In effect, these buildings suffered a double resonance condition: the amplification of the bedrock motion by the soil deposit and the amplification of the soil motion by the structure.

During the 1989 Loma Prieta earthquake, the San Francisco Bay Area (some 100 km north of the epicenter) experienced more damage than the epicentral region. The reason for the extensive damage along the margins of the bay is because of the amplification of the PGA at soft-soil sites. The bay basin is filled with an alluvial soil deposit of silts and clays, along with some layers of sandy and gravelly soils. The upper deposit, or young mud, is composed of silty clay, known as San Francisco Bay Mud. This material is of loose-to-medium density, and very compressible. As shown in Figure 2.9, the ground stratum in the vicinity of the bay is divided into three zones based on underlying material seismic response, from soft mud to hard rock. From Figure 2.9, it is clear that the attenuation of seismic waves varied considerably in different regions of the bay due to the varying subsurface soil conditions. The rock material attenuated the seismic waves relatively quickly compared to the soft young bay mud found at the margins of the bay. The areas underlain by this bay mud experienced the most damage because the soft mud greatly amplified ground motions.

These two examples clearly show that local site characteristics, particularly different underlying ground types, greatly influence the seismic wave frequency content and amplitude, as will be shown in Chapter 5. The National Earthquake Hazards Reduction Program (NEHRP) defines six ground types based on shear-wave velocity in order to establish amplification effects for design purposes:

1. Type A, hard rock (igneous rock)
2. Type B, rock (volcanic rock)
3. Type C, very dense soil and soft rock (sandstone)
4. Type D, stiff soil (mud)—it is conservative to assume this when no soil properties are known
5. Type E, soft soil (artificial fill)
6. Type F, soils requiring site-specific evaluations

Type A has the largest stiffness and generates the smallest amplifications, while Type E is the softest and generates the largest amplifications.

2.3.3 STRUCTURAL CHARACTERISTICS

Once seismic waves have been modified for ground conditions at the site of a structure, the structure responds to ground excitation based on its inherent characteristics, as well as the age of the structure and the quality of its construction. The characteristics of a structure that have the largest influence on its response include its weight and stiffness, and they will be covered in later chapters. Damping can play a significant role when structures remain within the elastic limit, but remains a much smaller effect for structures that deform into inelastic behavior as will be covered in Chapter 5. Although all these characteristics affect the displacement amplitude and natural period of a structure, stiffness has the greatest effect on structural response. A stiff structure vibrates with the same frequency and amplitude as the input motion, while a flexible structure may or may not vibrate in sync with the input motion, depending on how similar its natural period is to the input excitation period. For instance, long-period (low-frequency) input ground excitations amplify the deformations of flexible structures more than those of stiff ones, whereas short-period (high-frequency) input ground excitations amplify deformations of stiff structures more than those of soft ones. This structural response will be discussed in detail in later chapters.

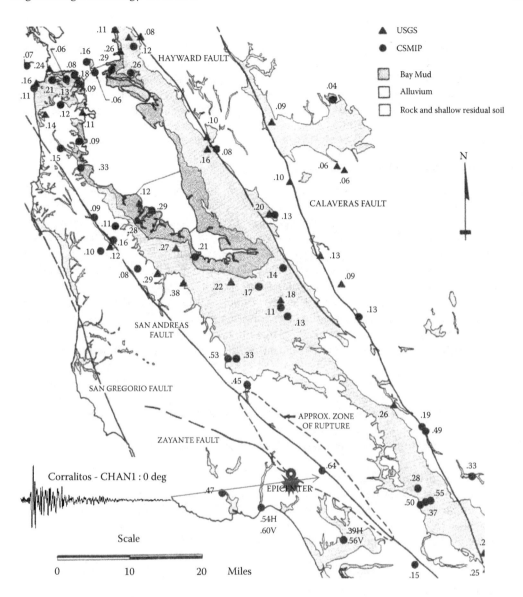

FIGURE 2.9 Horizontal PGA for the 1989 Loma Prieta earthquake (CSMIP = California Strong Motion Instrumentation Program). (Reproduced from Chopra, A. K., *Dynamics of Structures: Theory and Applications to Earthquake Engineering*, 4th edition, Prentice-Hall, Upper Saddle River, NJ, 2012. By permission of Prentice-Hall, Upper Saddle River, NJ.)

The age and construction of structures play a major role in their performance during an earthquake. Older and/or poorly constructed structures are at a greater risk of failure during an earthquake because of inadequate *ductility*, which is the ability of a structure to absorb the seismic energy without failure. A structure can include seismic detailing to satisfy the ductility demand during an earthquake, and to provide an adequate path for the seismic load to the foundation. Addressing ductility demand and load path through proper detailing can allow a structure to perform adequately during even major earthquakes. The ductility demand can be addressed by allowing the structure to behave in the inelastic range, provided that the stability of the overall system is not compromised. An alternative approach to resisting seismic loads is to permit the movement of structures; these systems are known as compliant systems.

Seismic loads on structures result from inertial forces created by the ground accelerations. Based on Newton's second law (force equals mass times acceleration), the magnitude of these forces is a function of the structure's mass (or weight). Also, the acceleration is a function of the characteristics of the input ground motion, particularly intensity, duration, and frequency content. The hazard a given earthquake risk possesses to structures of different functions is accounted for by specifying higher loads for structures of increased importance. Importance is classified in a range from unoccupied buildings to critical structures such as hospitals and schools; as discussed in Section 8.4.2.

2.4   EARTHQUAKE HAZARD ASSESSMENT

The intensity and magnitude scales are simplified methods for reporting the relative strength of an earthquake; however, these quantities do not necessarily provide a measure of seismic hazard at a particular site because not all regions are susceptible to strong earthquakes (small earthquakes occur all the time along active faults and generally do not pose a hazard to structures or human life). Also, intensity and magnitude are the measurements of past earthquakes and might not properly represent future ones. However, past regional seismicity (rate of occurrence of earthquakes) can be used to establish active seismic faults and forecast future seismic activity at a site. There are two quantitative methods that have been developed to characterize seismic hazard at a particular site: deterministic, where a particular earthquake is assumed, and probabilistic, where statistical methods are explicitly used to account for uncertainties in earthquake strength, location, and time of occurrence.

Even in highly seismic regions, the movement of the ground during an earthquake does not present a significant hazard to people (as discussed in Chapter 1, secondary effects such as fires, landslides, and tsunamis can). The highest risk to people in seismically active regions is shaking of buildings and other structures, particularly when structures collapse. The two methods used to characterize seismic hazards (deterministic and probabilistic) primarily focus on shaking effects.

The deterministic approach can be used in areas where seismic activity is frequent and its sources are well defined. The process is relatively simple and entails the following:

1. Identifying nearby seismic sources
2. Identifying the distance to the structure site from nearby seismic sources
3. Determining the magnitude and characteristics of nearby seismic sources
4. Establishing the structural response to the effects from all nearby seismic sources
5. Selecting the case that produces the largest structural response.

This method is too conservative because it does not properly account for uncertainties. For this reason, the probabilistic method was developed.

The probabilistic approach is similar, but includes the uncertainties in each step. The probabilistic approach, although quite complicated, has been incorporated in the various seismic hazard maps used in contemporary seismic design codes. These codes have incorporated probabilistic seismic hazard zone maps based on past geographic seismology to minimize risk to human life. These maps depict where damage might occur, based on the probability of the occurrence of strong shaking and the local geology (areas underlain by deep, loose sediments are at a greater risk for damage than those directly over solid rock); see Figure 2.10. These maps can be used to draw probabilistic design spectra, which will be discussed in detail in Chapters 5 and 8. With this information, structural engineers can plan and design structures at a particular site using the methods covered in Chapter 8.

The key issue in the probabilistic approach found in the seismic design codes is the probability of exceedance of a particular event for a specified intensity, over a certain period of time. The codes consider three different earthquake levels for design: the maximum design earthquake (MDE), which has a 10% probability of being exceeded over a period of 50 years; one with 5% probability of being exceeded over a period of 50 years; and the maximum considered earthquake (MCE), which has a 2% probability of being exceeded over a period of 50 years. Also, an earthquake

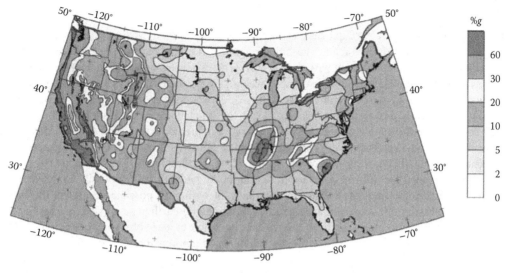

Horizontal ground acceleration (%g)
with 2% probability of exceedance in 50 years

FIGURE 2.10 Probabilistic seismic hazard map showing ground shaking hazard zones. (From USGS National Seismic Hazard Mapping Project website: geohazards.cr.usgs.gov.)

that has a significant chance of occurring during the life of a structure is known as the maximum probable earthquake (MPE), and has a 50% probability of being exceeded over a period of 50 years. Designing for the MCE provides the highest level of protection and is reserved only for the most important or critical structures. These earthquakes can also be characterized in terms of the return period (RP) using the following relationship:

$$RP = \frac{-T}{\ln(1-P)} c \tag{2.8}$$

where:

P is the probability of exceedance in T years

That is, the RPs in years for the three earthquake levels included in codes and the MPE are as follows:

$$MCE \rightarrow RP = \frac{-T}{\ln(1-P)} = \frac{-50 \text{ years}}{\ln(1-0.02)} = 2475 \text{ years}$$

$$RP = \frac{-T}{\ln(1-P)} = \frac{-50 \text{ years}}{\ln(1-0.05)} = 975 \text{ years}$$

$$MDE \rightarrow RP = \frac{-T}{\ln(1-P)} = \frac{-50 \text{ years}}{\ln(1-0.10)} = 475 \text{ years}$$

$$MPE \rightarrow RP = \frac{-T}{\ln(1-P)} = \frac{-50 \text{ years}}{\ln(1-0.50)} = 72 \text{ years}$$

2.5   EARTHQUAKE PREDICTION

The temporal prediction of earthquakes has been investigated for many years without success. Also, all signs, such as radon detection and unusual animal behavior, that have been claimed as portending earthquakes have been proven wrong. However, sensitive instruments can detect the arrival of P-waves that are followed by the devastating S-waves. This is the current basis for proposed early earthquake detection systems. In some parts of the world, these systems have already been deployed and have successfully provided warnings to residents of afflicted areas, albeit short warnings.

Although it is unlikely that we will develop the ability to predict *when* earthquakes will occur, it is generally possible to predict *where* earthquakes can occur. Great advances have been made in the past century to map the most seismic-prone regions of the world, since past seismic activity in these regions has provided the information needed to map many faults. It is now possible to predict the spatial occurrence of earthquakes relatively accurately, including the possible strength of the event, as well as forecasting the expected average rates of damaging earthquakes. Also, because of the instrumental records from past earthquakes, we have been able to map subsurface conditions, which have allowed us to predict the characteristics of the ground motions that may occur at a particular site from a statistical stand point.

PROBLEMS

2.1   A Richter magnitude earthquake of 6.8 would radiate how much more energy than an M_L 5.8 earthquake?

2.2   A Richter magnitude earthquake of 5.0 is how much stronger than an M_L 3.0 earthquake?

2.3   Estimate the local magnitude of a southern California earthquake recorded using a standard Wood–Anderson seismograph that shows a trace amplitude of 23 mm, and the P- and S-wave arrival times at the recording station of 07:19:45 and 07:20:09, respectively.

2.4   An earthquake at a transformed fault caused an average strike-slip displacement of 2.5 m over an area equal to 80 km long by 23 km deep. Assuming the rock along the fault has the average shear stiffness of 175 kPa, estimate the seismic moment and moment magnitude of the earthquake.

2.5   Estimate the seismic moment and moment magnitude of the February 27, 2010 Chile earthquake. It is estimated that the thrust fault (the slip plane ends before reaching the earth's surface) caused an average strike-slip displacement of 5 m over an area equal to 600 km long by 150 km deep. Assume the rock along the fault has the average shear rigidity of 3×10^{11} dyne/cm^2.

2.6   What is the energy difference between the Haiti and Chile earthquakes from Example 2.2 and Problem 2.5, respectively?

REFERENCES

Bolt, B. A., *Earthquakes*, 5th edition, W.H. Freeman and Co., New York, 2004.

Chopra, A. K., *Dynamics of Structures: Theory and Applications to Earthquake Engineering*, 4th edition, Prentice-Hall, Upper Saddle River, NJ, 2012.

Eberhard, M. O., S. Baldridge, J. Marshall, W. Mooney, and G. J. Rix, *The MW 7.0 Haiti earthquake of January 12*, 2010; USGS/EERI Advance Reconnaissance Team Report: U.S. Geological Survey Open-File Report 2010-1048, 58 p, 2010.

Kramer, S. L., *Geotechnical Earthquake Engineering*, Prentice-Hall, Upper Saddle River, NJ, 1996.

Richter, C. F., *Elementary Seismology*, W.H. Freeman and Co., San Francisco, CA, 1958.

Villaverde, R., *Fundamental Concepts of Earthquake Engineering*, CRC Press, Boca Raton, FL, 2009.

3 Single-Degree-of-Freedom Structural Dynamic Analysis

After reading this chapter, you will be able to:

1. Determine the weight of a structure
2. Determine the stiffness of a structure
3. Determine the natural angular frequency of a structure
4. Determine the natural period of a structure
5. Determine the natural cyclic frequency of a structure
6. Perform a dynamic analyses of damped and undamped single-degree-of-freedom systems
7. Perform a dynamic analyses of damped and undamped single-degree-of-freedom systems subjected to harmonic time-dependent forces
8. Determine deflections and stresses induced in a structure by motions of its supports

Traditional seismic design follows two approaches: one based on static analysis and the other one based on dynamic analysis (see Chapter 8). The static analysis is formulated by estimating an equivalent lateral seismic load as a fraction of the total system weight; the fraction is determined from a notional earthquake, which is based on the risk posed by the seismicity of the site and is captured in a single graph known as the response spectrum (to be discussed in Chapters 4 and 5). The dynamic analysis procedure models the load effect of earthquakes more accurately by using either a set of specific earthquake ground motion time-history records (response time-history analysis procedure) or the effect of the same notional earthquake as the equivalent static analysis (the design response spectrum), but using a modal analysis (modal response spectrum analysis procedure). In the response time-history analysis procedure, a linear or nonlinear mathematical model of the system is used to determine the system response (displacements and accelerations, which are used to determine the internal loading in each member) at each increment of time for a suite of ground motion acceleration time histories. With the full history of the response, the absolute maximum internal loads can be determined by combining the results of the set of ground motion acceleration time histories; these loads can then be used to design each member. The modal response spectrum analysis procedure uses established methods of structural dynamics to determine system vibration mode shapes and their associated natural periods, which along with the response spectrum are used to establish the maximum structural response. With this maximum response, the maximum internal loads can be determined and used to design each member. In this chapter, we explain the essential concepts of structural dynamics, also known as vibration theory, needed to understand both dynamic seismic analyses procedures.

3.1 UNDAMPED SINGLE-DEGREE-OF-FREEDOM SYSTEM

The concept of degree-of-freedom (DOF) is central to the understanding of vibration theory. A DOF in the context of vibration theory is an independent displacement or rotation of a point on a structure, which means that even the simplest case could potentially have an infinite number of DOFs. However, determination of the number of DOFs depends on the idealization of the structural system. The ends of a member are often points of interest; for example, consider a portal frame with rigid connections between columns and beam and fixed supports at the base of each column. The ends of the column at the supports would have zero DOFs, but three DOFs at each of the connections between beam and column (see Figure 3.1). These connections must both move and rotate in

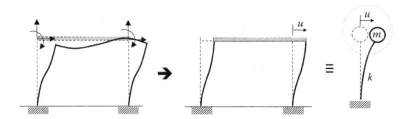

FIGURE 3.1 Idealized single-degree-of-freedom system for a portal frame.

sync, resulting in six DOFs. However, for our discussion in this chapter, the beam will be assumed to be much stiffer than the columns, preventing rotations of the connections. Also, the two columns move in parallel, and if we assume that the axial deformations of the columns and beam are relatively small, we can reduce the entire system to a single-degree-of-freedom (SDOF) case as shown in Figure 3.1. Finally, note that there is no damping in the system, the lack of which would cause the system to move in perpetuity once set in motion. We will cover damping in Section 3.2.

3.1.1 FREE VIBRATION RESPONSE OF UNDAMPED SYSTEMS

Free vibration is caused by initial conditions (displacement and velocity) and not by applied force. To characterize the free vibration motion of the portal frame idealized as an SDOF system, we first need to formulate a mathematical model in terms of the displacement as a function of time. This can be accomplished by establishing an equation of motion for the system. This equation can be determined by applying equilibrium to a free-body diagram (FBD) of the mass using D'Alembert's principle, where the inertial force developed by the mass (from Newton's second law, $F = ma$) is included in the FBD as shown in Figure 3.2. Note that this is not a complete FBD because the internal axial loading and the internal moment in the column stem are missing.

Horizontal equilibrium of the FBD shown in Figure 3.2 yields the equation of motion,

$$+ \rightarrow \sum F_x = 0; \quad -m\ddot{u} - ku = 0 \Rightarrow m\ddot{u} + ku = 0$$

where:

m is the mass of the system, discussed in more detail in Section 3.1.2
k is the lateral stiffness, discussed in more detail in Section 3.1.3

The double dot over the u indicates differentiation with respect to time, that is,

$$\ddot{u} = \frac{d^2 u}{dt^2}$$

The equation of motion, a second-order, linear, and homogeneous differential equation with constant coefficients, can be written in the following form:

$$\ddot{u} + \omega_n^2 u = 0 \tag{3.1}$$

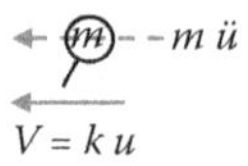

$$\leftarrow \ \widehat{m} \ \text{-}\,\text{-}\, m\ddot{u}$$
$$\overline{V = ku}$$

FIGURE 3.2 Free-body diagram of a portal frame SDOF model.

where for convenience a new parameter is introduced, the natural circular frequency (from now on called natural frequency) with units of radians per second (rad/s),

$$\omega_n = \sqrt{\frac{k}{m}} \tag{3.2}$$

The solution to the second-order, linear, and homogeneous differential equation of motion is of the form

$$u(t) = e^{\lambda t}$$

This equation can be differentiated with respect to time to obtain the velocity,

$$\dot{u}(t) = \lambda e^{\lambda t}$$

and the velocity can be differentiated with respect to time to obtain the acceleration,

$$\ddot{u}(t) = \lambda^2 e^{\lambda t}$$

substituting the equations for displacement and acceleration into the equation of motion yields

$$\lambda^2 e^{\lambda t} + \omega_n^2 e^{\lambda t} = 0$$

which can be rearranged as

$$\left(\lambda^2 + \omega_n^2\right) e^{\lambda t} = 0$$

In general, the exponential function is not zero, thus the quantity within parenthesis must be zero in order to determine a valid solution to the differential equation,

$$\left(\lambda^2 + \omega_n^2\right) = 0$$

We then solve for the unknown parameter λ,

$$\lambda = \pm \omega_n i$$

where i is the imaginary unit of a complex number given as $i = \sqrt{-1}$. As the equation of motion is a second-order differential equation, two constants of integration are needed, A_1 and A_2,

$$u(t) = A_1 e^{\omega_n i t} + A_2 e^{-\omega_n i t}$$

This equation can be expressed in a polar form (in terms of sines and cosines) by making use of Euler's identities:

$$e^{i\theta} = \cos\theta + i\sin\theta \quad \text{and} \quad e^{-i\theta} = \cos\theta - i\sin\theta$$

The result is

$$u(t) - A_1(\cos \omega_n t + i \sin \omega_n t) + A_2(\cos \omega_n t - i \sin \omega_n t)$$

which can be rewritten as

$$u(t) = (A_1 + A_2)\cos \omega_n t + i(A_1 - A_2)\sin \omega_n t$$

As the two trigonometric functions are real-valued solutions to the equation of motion, we can express the general solution in real form as

$$u(t) = A \cos \omega_n t + B \sin \omega_n t \tag{3.3}$$

where A and B are arbitrary constants.

We now use initial conditions (at time $t = 0$) to solve for constants A and B. First, we substitute the initial displacement (at time $t = 0$), $u(0)$, into Equation 3.3 in order to determine the constant A,

$$u(0) = A\cos(0) + B\sin(0)$$

which yields $A = u(0)$. Next, we take a time derivative of Equation 3.3 and substitute the initial velocity (at time $t = 0$), $\dot{u}(0)$, in order to determine the constant B,

$$\dot{u}(0) = -A\omega_n \sin(0) + B\omega_n \cos(0)$$

which yields $B = (\dot{u}(0)/\omega_n)$.

The complete solution is the free vibration of an undamped SDOF system, which describes the position of the mass as a function of time,

$$u(t) = u(0)\cos \omega_n t + \frac{\dot{u}(0)}{\omega_n}\sin \omega_n t \tag{3.4}$$

A graph of the solution is shown in Figure 3.3. This motion is described as harmonic (and therefore periodic) because it is a function of sine and cosine of the same frequency, ω_n. The time required to complete a full cycle (2π) is known as the natural period of vibration, T_n (s), and is determined as (see Figure 3.3)

$$\omega_n T_n = 2\pi \Rightarrow T_n = \frac{2\pi}{\omega_n} \tag{3.5}$$

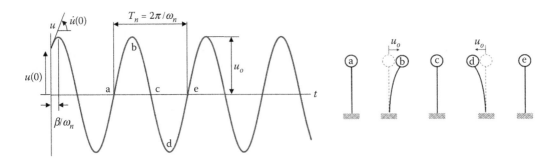

FIGURE 3.3   Free vibration of an undamped SDOF system.

The natural cyclic frequency, f_n (hertz or cycles/s), is defined as the reciprocal of the natural period of vibration, T_n (s), and is proportional to the natural circular frequency, ω_n (rad/s), that is,

$$f_n = \frac{1}{T_n} = \frac{\omega_n}{2\pi} \tag{3.6}$$

Note that these three quantities are related and essentially represent the same physical quantity expressed in different units.

An alternate formulation for the free vibration response of an undamped SDOF leads to the definition of two additional parameters introduced in Figure 3.3: the phase angle, β, and the amplitude, u_o. We start with Equation 3.3 and assume that constants A and B are related to new variables C and β as follows:

$$A = C\cos\beta \quad \text{and} \quad B = C\sin\beta \tag{3.7}$$

Squaring these two relationships and adding the results yields

$$A^2 + B^2 = C^2(\cos^2\beta + \sin^2\beta)$$

The quantity in parentheses is equal to 1.0 (trigonometric identity); thus,

$$C = \sqrt{A^2 + B^2} = \sqrt{u(0)^2 + (\dot{u}(0)/\omega_n)^2} = u_o$$

where u_o is the maximum amplitude shown in Figure 3.3.

Also, taking the ratio of B to A, we obtain the phase angle, β,

$$\frac{B}{A} = \frac{C\sin\beta}{C\cos\beta} = \tan\beta$$

or

$$\beta = \tan^{-1}\left(\frac{B}{A}\right) = \tan^{-1}\left(\frac{\dot{u}(0)}{u(0)\omega_n}\right)$$

Substituting Equation 3.7 into Equation 3.3 yields the equation of motion in terms of the amplitude and phase angle, that is,

$$u(t) = A\cos\omega_n t + B\sin\omega_n t = C\cos\beta\cos\omega_n t + C\sin\beta\sin\omega_n t$$

which can be rearranged using the following trigonometric identities:

$$\cos\beta\cos\omega t = \frac{1}{2}\cos(\omega t - \beta) + \frac{1}{2}\cos(\omega t + \beta)$$

$$\sin\beta\sin\omega t = \frac{1}{2}\cos(\omega t - \beta) - \frac{1}{2}\cos(\omega t + \beta)$$

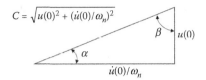

FIGURE 3.4 Relationship between phase angles and amplitude.

The equation in the new form is

$$u(t) = C\cos(\omega_n t - \beta)$$

Similarly,

$$u(t) = C\sin(\omega_n t + \alpha)$$

where a new phase angle, α, is introduced,

$$\alpha = \tan^{-1}\left(\frac{u(0)\omega_n}{\dot{u}(0)}\right)$$

The two phase angles and the amplitude are related as shown in Figure 3.4.

3.1.2 STRUCTURAL WEIGHT

As noted earlier, the circular frequency is proportional to the mass, which in turn is directly proportional to the weight. For seismic analysis of structures, the weight is defined as the total effective weight W, which is specified in ASCE/SEI (2010) "Minimum Design Loads for Buildings and Other Structures" standard, (from here on referred to as ASCE-7) as

$$W = DL + 0.25StL + \text{larger of } PL \text{ or } 10\text{-psf} + WPE + 0.2SL + LaL \qquad (3.8)$$

where:
  DL is the dead load of the structural system that is tributary to each floor
  StL is the storage load, which is a live load in areas used for storage, such as a warehouse
  PL is the partition load, when applicable
  WPE is the operating weight of permanent equipment
  SL is the flat roof snow load when it exceeds 30 psf, regardless of roof slope
  LaL is the landscape loads associated with roof and balcony gardens

For partition loading, ASCE-7 requires the given partition load or 10 psf when determining the effective weight. This is only applicable where the location of partitions is subject to change (such as buildings designated as "office building" occupancy), where ASCE-7 requires a partition live load of 15 psf for designing individual floor members for vertical loads, if the floor live load is <80 psf. For information about estimating the other loads, see ASCE-7 specifications.

Once the dead load is estimated, one can determine the tributary weight of a floor or roof as shown in Figure 3.5 for a multistory building. The weight of each floor and the tributary wall loads halfway between adjacent floors is assumed to be concentrated or lumped at each floor level (see Figure 3.6). The story weight, W_x, for this diagram is given as

$$W_x = \text{walls (A, B, C)} + \text{floor (D)} + \text{equipment (E)} = W_A + W_B + W_C + W_D + W_E$$

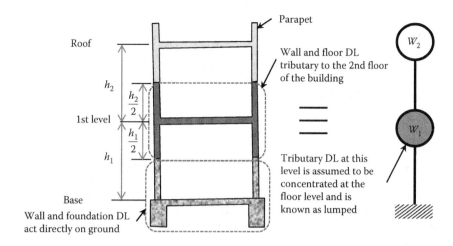

FIGURE 3.5   Building structure and lumped mass simplified model.

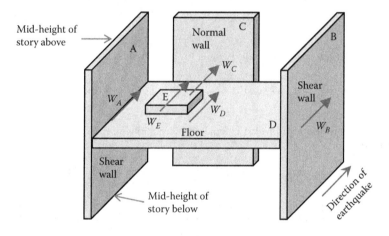

FIGURE 3.6   Story weight, W_x for calculating lateral forces.

Here W_D includes weight of the structure floor, suspended ceiling, mechanical equipment (unless taken separately as W_E), and when applicable 10 psf for partitions or the actual partition weight. In general, foundation weight and half of the first-story wall weight are assumed to act directly on the ground and are commonly omitted in seismic calculations (though common practice, this is not an explicit provision in seismic design specifications).

EXAMPLE 3.1

The load pressures for dead, live, and snow loads for the rectangular (30 ft × 40 ft) warehouse structure shown are provided below. Determine the effective weight at the roof.

Roof DL = 45 psf
Floor DL = 65 psf
Wall DL = 25 psf
Floor LL = 250 psf (storage)
SL = 50 psf

SOLUTION

1. Identify the effective height of the roof in order to determine the tributary height of the walls needed to calculate the effective weight of the walls.

   In this case, the tributary height of the walls is the sum of the height from the roof to midway to the base floor and the entire height of a parapet wall above the roof level as shown in Figure E3.1:

$$h_{tributary} = \frac{h_1}{2} + h_{parapet} = \frac{16\ ft}{2} + 4\ ft = 12\ ft$$

2. Calculate individual weights of components associated with the floor of interest.

   The components contributing to the effective weight of the roof include the walls parallel to north–south (N–S) and east–west (E–W) directions, roof load, and snow load. Note that the storage live load and floor dead load do not contribute to the seismic weight of the roof because they are associated with the base level and act directly on the ground.

   The weight of the walls is calculated by multiplying the area of the walls with the wall dead load.

   Weight of walls parallel to the N–S direction:

$$W_{N-S,Walls} = Area_{N-S,Walls} \times Wall\,DL = (2\ walls)\frac{(40\ ft \times 12\ ft)}{wall} \times 25\ psf = 24{,}000\ lb$$

   Similarly, for the weight of walls parallel to the E–W direction:

$$W_{E-W,Walls} = (2\ walls)\frac{(30\ ft \times 12\ ft)}{wall} \times 25\ psf = 18{,}000\ lb$$

   The weight of the roof level is calculated by multiplying the roof area with the roof dead load.

$$W_{Roof\ Level} = Area_{Roof\ Level} \times Roof\ DL = (30\ ft \times 40\ ft) \times 45\ psf = 54{,}000\ lb$$

   This problem specifies a snow load pressure which is necessary for the seismic weight calculation of the roof level. Based on Equation 3.7, snow loads exceeding 30 psf are factored by 0.2 in the seismic weight calculation.

$$W_{Snow} = Area_{Roof\ Level} \times (0.2 \times SL) = (30\ ft \times 40\ ft) \times (0.2 \times 50\ psf) = 12{,}000\ lb$$

3. The effective seismic weight for the roof can be determined by summing the weight contribution from each of the components.

   As the roof is the only level above the base, the effective seismic weight of the roof is denoted as W_1:

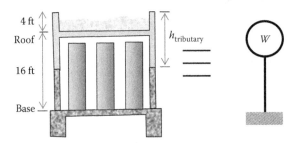

FIGURE E3.1 One story schematic for effective weight calculations for an idealized structural model.

$$W_1 = W_{\text{N-S,Walls}} + W_{\text{E-W,Walls}} + W_{\text{Roof Floor}} + W_{\text{Snow}}$$

$$W_1 = 24{,}000\ \text{lb} + 18{,}000\ \text{lb} + 54{,}000\ \text{lb} + 12{,}000\ \text{lb} = 108{,}000\ \text{lb} = 108\ \text{kips}$$

EXAMPLE 3.2

Given the following rectangular (30 ft × 40 ft) two-story warehouse and loading shown, determine the effective weight at the roof and first floor:

Roof DL = 45 psf
Floor DL = 65 psf
Wall DL = 25 psf
Floor LL = 250 psf (storage)
SL = 50 psf

SOLUTION

1. Identify the effective height of the first floor and the roof in order to calculate the effective weight of the walls at each level.

   The effective height of the first level is the sum of the heights halfway to the adjacent levels as shown in Figure E3.2:

$$h_{\text{effective},1} = \frac{h_1}{2} + \frac{h_2}{2} = \frac{16\ \text{ft}}{2} + \frac{14\ \text{ft}}{2} = 15\ \text{ft}$$

   The effective height of the roof is the sum of the heights from the roof midway to the first level and the entire height of the parapet wall above the roof floor as shown in Figure E3.2:

$$h_{\text{effective,roof}} = \frac{h_2}{2} + h_{\text{parapet}} = \frac{14\ \text{ft}}{2} + 2\ \text{ft} = 9\ \text{ft}$$

2. Calculate individual weights of components associated with each level of interest.

   First level component weights:

   For the first level, the seismic weight includes dead loads from the walls and floor, whereas the live load is a result of storage load.

   Weight of walls parallel to the N–S direction,

$$W_{\text{N-S,Walls},1} = \text{Area}_{\text{N-S,Walls},1} \times \text{Wall DL} = (2\,\text{walls}) \frac{(40\ \text{ft} \times 15\ \text{ft})}{\text{wall}} \times 25\ \text{psf} = 30{,}000\ \text{lb}$$

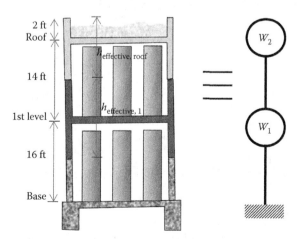

FIGURE E3.2 Two story schematic for effective weight calculations for an idealized structural model.

Similarly, for the weight of walls parallel to the E–W direction,

$$W_{E-W,Walls,1} = (2 \text{ walls})\frac{(30 \text{ ft} \times 15 \text{ ft})}{\text{wall}} \times 25 \text{ psf} = 22,500 \text{ lb}$$

The floor load components on the first level include the floor dead load and the storage live load,

$$W_{Floor\ DL,1} = Area_{Floor,1} \times FloorDL = (30 \text{ ft} \times 40 \text{ ft}) \times 65 \text{ psf} = 78,000 \text{ lb}$$

This problem specifies a storage live load pressure which is necessary for the seismic weight calculation of the first level. Based on Equation 3.7, storage live load, *StL*, is factored by 0.25 in the seismic weight calculation.

$$W_{Floor\ LL,1} = Area_{Floor,1} \times 0.25 \times StL = (30 \text{ ft} \times 40 \text{ ft}) \times 0.25 \times 250 \text{ psf} = 75,000 \text{ lb}$$

Roof level component weights:
For the roof level, the seismic weight includes dead loads from the walls and floor and the snow load. The process for calculating the effective seismic weight is detailed in Example 3.1.
Weight of walls parallel to the N–S direction,

$$W_{N-S,Walls,roof} = (2 \text{ walls})\frac{(40 \text{ ft} \times 9 \text{ ft})}{\text{wall}} \times 25 \text{ psf} = 18,000 \text{ lb}$$

Weight of walls parallel to the E–W direction,

$$W_{E-W,Walls,roof} = (2 \text{ walls})\frac{(30 \text{ ft} \times 9 \text{ ft})}{\text{wall}} \times 25 \text{ psf} = 13,500 \text{ lb}$$

The roof floor load components include the floor dead load and the snow load,

$$W_{Floor\ DL,\ roof} = Area_{Roof\ Floor} \times Roof\ DL = (30 \text{ ft} \times 40 \text{ ft}) \times 45 \text{ psf} = 54,000 \text{ lb}$$

$$W_{Snow} = Area_{Roof\ Floor} \times (0.2 \times SL) = (30 \text{ ft} \times 40 \text{ ft}) \times (0.2 \times 50 \text{ psf}) = 12,000 \text{ lb}$$

3. Determine the effective seismic weight for each level.
   The effective seismic weight for each level can be determined by summing the weight of the components. Level 1 is denoted as W_1 and the roof level is denoted as W_2:

$$W_1 = W_{N-S,Walls,1} + W_{E-W,Walls,1} + W_{Floor\ DL,1} + W_{Floor\ LL,1}$$

$$W_1 = 30,000 \text{ lb} + 22,500 \text{ lb} + 78,000 \text{ lb} + 75,000 \text{ lb} = 205,500 \text{ lb} = 205.5 \text{ kips}$$

$$W_2 = W_{Roof} = W_{N-S,Walls,Roof} + W_{E-W,Walls,Roof} + W_{Floor\ DL,Roof} + W_{Snow}$$

$$W_2 = 18,000 \text{ lb} + 13,500 \text{ lb} + 54,000 \text{ lb} + 12,000 \text{ lb} = 97,500 \text{ lb} = 97.5 \text{ kips}$$

3.1.3 STRUCTURAL STIFFNESS

To completely characterize the equation of motion for a linear elastic SDOF system, we need the lateral stiffness, *k*. This variable, along with the mass, is needed to determine the circular frequency, which once obtained can be used to determine the natural period of vibration of the system. The stiffness is defined as the force or moment that results in a unit displacement or rotation at a DOF.

EXAMPLE 3.3

Determine the stiffness for a simply supported beam with a concentrated load F applied at mid-span. This loading can be considered lateral with respect to the axis of bending.

SOLUTION

1. Determine the deflection at point of interest (or DOF) for the given loading.
   The deflection of the beam can be determined with one of the many structural analyses methods (e.g., double integration, conjugate beam, virtual work, and moment-area) used to compute deflections. Also, beam deflections can be obtained from deflection tables such as those found in Part 4 of the American Institute of Steel Construction Manual. The midspan deflection for a simply supported beam with a concentrated load F is shown in Figure E3.3.
2. Determine the stiffness for the given loading and associated DOF.
   The stiffness k is the ratio of force (or moment) to displacement (or rotation). The midspan deflection can be rearranged to determine the stiffness,

$$k = \frac{48EI}{L^3}$$

where the quantity EI is the flexural stiffness given by the product of the modulus of elasticity, E, and the second moment of the cross-section about the axis of bending (moment of inertia), I.

Following the approach of Example 3.3, we can derive equations for k for a number of simple structural systems that can be modeled as SDOF (see Table 3.1 for a list of cases).

EXAMPLE 3.4

A fixed–fixed beam of length, L, with a concentrated weight, W, applied at midspan has a constant flexural rigidity, EI. Idealize the beam and weight shown in Figure E3.4 into an SDOF system and determine its natural and cyclic frequencies.

SOLUTION

1. Idealize the structural system.
   In order to idealize or model this structural system as a SDOF system, we assume that the weight of the beam is small relative to the concentrated weight, W. In addition, we limit the motion of the weight to vertical displacements, u, resulting in the lumped mass model shown in Figure E3.5.
2. Determine the stiffness parameters.
   Considering the displaced shape of the SDOF system as shown in Figure E3.6, the static load–displacement relationship can be used to determine the stiffness associated with the DOF.

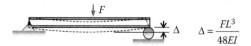

FIGURE E3.3 Maximum deflection of a simply supported beam with concentrated force at midspan.

TABLE 3.1
Equivalent Stiffness Constants, k

Case	Max Deflection, Δ	Stiffness, k
Case 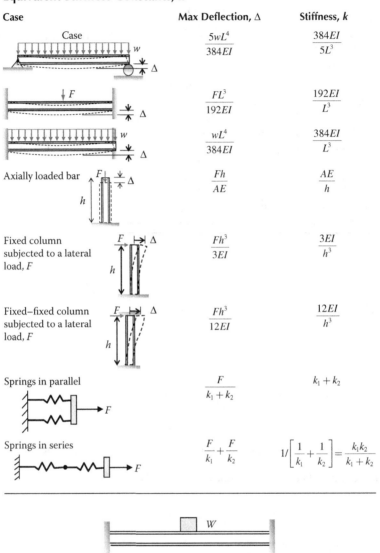	$\dfrac{5wL^4}{384EI}$	$\dfrac{384EI}{5L^3}$
	$\dfrac{FL^3}{192EI}$	$\dfrac{192EI}{L^3}$
	$\dfrac{wL^4}{384EI}$	$\dfrac{384EI}{L^3}$
Axially loaded bar	$\dfrac{Fh}{AE}$	$\dfrac{AE}{h}$
Fixed column subjected to a lateral load, F	$\dfrac{Fh^3}{3EI}$	$\dfrac{3EI}{h^3}$
Fixed–fixed column subjected to a lateral load, F	$\dfrac{Fh^3}{12EI}$	$\dfrac{12EI}{h^3}$
Springs in parallel	$\dfrac{F}{k_1+k_2}$	k_1+k_2
Springs in series	$\dfrac{F}{k_1}+\dfrac{F}{k_2}$	$1/\left[\dfrac{1}{k_1}+\dfrac{1}{k_2}\right]=\dfrac{k_1 k_2}{k_1+k_2}$

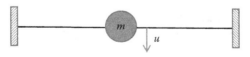

FIGURE E3.4 Fixed–fixed beam with concentrated weight at midspan.

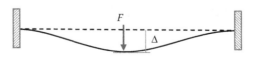

FIGURE E3.5 Idealized lumped mass model of a fixed–fixed beam with concentrated weight at midspan.

FIGURE E3.6 Displaced shape of a fixed–fixed beam with concentrated force at midspan.

From Table 3.1, the force–displacement relationship is given by

$$\Delta = \frac{FL^3}{192EI}$$

The stiffness of the SDOF system is determined by applying the same process as in Example 3.3:

$$k = \frac{F}{\Delta} = \frac{192EI}{L^3}$$

3. Determine the natural and cyclic frequencies.

The equation of motion for this SDOF system under free vibration is described by Equation 3.1:

$$\ddot{u} + \omega_n^2 u = 0$$

where the natural frequency is determined using Equation 3.2. Substituting for stiffness and mass in terms of known quantities yields

$$\omega_n = \sqrt{\frac{192(EI/L^3)}{(W/g)}} = 8\sqrt{\frac{3EIg}{WL^3}} \text{ rad/s}$$

The natural frequency, ω_n, results in units of radians per second. In order to determine the cyclic frequency, f, the units are converted from radians per second into cycles per second by applying the conversion of 1 cycle $= 2\pi$ radians or using Equation 3.6:

$$f = \omega_n \left(\frac{1 \text{cycle}}{2\pi \text{rad}}\right) = \frac{4}{\pi}\sqrt{\frac{3EIg}{WL^3}} \text{ cycles/s} \quad \text{or} \quad \text{hertz}$$

EXAMPLE 3.5

Given a water tank supported on a slender column as shown in Figure E3.7, determine the natural frequency and period of the system.

SOLUTION

1. Idealize the structural system.

Similar to Example 3.4, the weight of the column is considered negligible with respect to the weight of the water tank. Considering only the horizontal displacement of the water tank, the lumped mass model is shown in Figure E3.7.

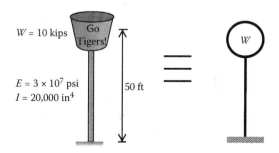

$W = 10$ kips

$E = 3 \times 10^7$ psi
$I = 20{,}000$ in^4

50 ft

W

FIGURE E3.7   Water tank supported on a slender column and an idealized structural model.

2. Determine the stiffness parameters.

From Table 3.1, the stiffness of the SDOF system is determined by applying the same process described in Examples 3.3 and 3.4:

$$k = \frac{3EI}{L^3} = \frac{3(3 \times 10^7 \text{ psi})(20,000 \text{ in}^4)}{(50 \text{ ft} \times 12 \text{ in/ft})^3} = 8333 \text{ lb/in}$$

3. Determine the natural frequency and period of the system.

As the water tank is modeled as an SDOF system, the equation of motion is described by Equation 3.1. The natural frequency of water tank is determined using Equation 3.2:

$$\omega_n = \sqrt{\frac{k}{m}} = \sqrt{\frac{8333 \text{ lb/in}}{(10,000 \text{ lb}/(32.2 \text{ ft/s}^2 \times 12 \text{ in/ft}))}} = 17.94 \text{ rad/s}$$

The natural period of vibration is determined using Equation 3.5:

$$T_n = \frac{2\pi}{\omega_n} = \frac{2\pi}{17.94 \text{ rad/s}} = 0.35 \text{ s}$$

EXAMPLE 3.6

The space-grid roof structure may be considered rigid, and it has a dead load of 20 psf. The side sheathing has a dead load of 10 psf. All steel columns are $W10 \times 30$ ($I = 170 \text{ in}^4$). Determine the dynamic properties, natural frequency, and period, in the E–W direction. Note that the first pair of columns in Figure E3.8 is fixed–pinned, the second is pinned–pinned, and the last pair is fixed–fixed. A plan view of the structure is shown in Figure E3.9.

SOLUTION

1. Idealize the structural system into an SDOF system.

The frame can be modeled as an SDOF system assuming an rigid roof and the total stiffness of the SDOF system is the sum of the individual column stiffnesses.

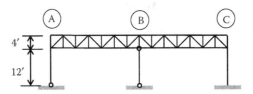

FIGURE E3.8   Elevation view of a space grid roof structure.

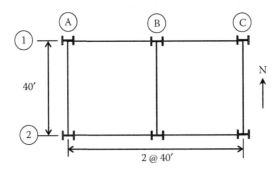

FIGURE E3.9   Plan view of a space grid roof structure.

2. Determine the weight (proportional to the mass) of the SDOF system.

The weight of the SDOF system includes the roof dead load of 20 psf and the wall sheathing weight around the perimeter of the building (240 ft). The weight due to sheathing includes the tributary height of the roof, 10 ft, in the area calculation:

$$W = 20\,\text{lb/ft}^2(40\,\text{ft} \times 80\,\text{ft}) + 10\,\text{lb/ft}^2(10\,\text{ft} \times 240\,\text{ft}) = 88{,}000\,\text{lb} = 88\,\text{kips}$$

3. Determine stiffness parameters.

The lateral stiffness of the columns depends on the boundary conditions and connections to the roof beams. In this case, columns A1 and A2 are pinned and fixed at their ends, columns B1 and B2 are pinned–pinned connected at their ends, and columns C1 and C2 are fixed–fixed connected at their ends.

A1 and A2 column stiffnesses,

$$k_{A1} = k_{A2} = \frac{3EI}{h^3}$$

B1 and B2 column stiffnesses,

$$k_{B1} = k_{B2} = 0$$

C1 and C2 column stiffnesses

$$k_{C1} = k_{C2} = \frac{12EI}{h^3}$$

The total stiffness of the SDOF system is the sum of column stiffnesses in the E–W direction.

$$K_{E-W} = 2 \times \frac{3EI}{h^3} + 2 \times \frac{12EI}{h^3} = \frac{30EI}{h^3} = \frac{30(29{,}000\,\text{ksi})(170\,\text{in}^4)}{(144\,\text{in})^3} = 49.5\,\text{kips/in}$$

4. Determine the natural frequency and period of the SDOF system.

As the frame is modeled as an SDOF system, the equation of motion is described by Equation 3.1. The natural frequency of the roof structure is determined using Equation 3.2:

$$\omega_n = \sqrt{\frac{K_{E-W}}{m}} = \sqrt{\frac{49.5\,\text{kips/in}}{(88\,\text{kips}/(32.2\,\text{ft/s}^2 \times 12\,\text{in/ft}))}} = 14.74\,\text{rad/s}$$

The natural period of vibration is determined using Equation 3.5:

$$T_n = \frac{2\pi}{\omega_n} = \frac{2\pi}{14.74\,\text{rad/s}} = 0.43\,\text{s}$$

Lateral force-resisting systems are used to resist lateral seismic (or wind) loads and are generally categorized into one of three different categories: *unbraced frames, braced frames,* or *shear walls.* Following are simplified derivations of the equivalent lateral stiffness constants for single-story systems of each of these three cases.

Unbraced frames have rigid connections between columns and beams and carry lateral load by developing moments in the columns and beams. If we consider the portal frame introduced earlier in this chapter (see Figure 3.1), we can derive the equivalent stiffness of the frame by assuming that each column contributes equally to the frame stiffness. For example, for a portal frame with fixed–fixed columns (such as the one listed in the sixth row of Table 3.1), which would be equivalent to a frame with a very stiff beam, the total stiffness is given as

$$K = (2)\frac{12EI}{h^3} = \frac{24EI}{h^3} \tag{3.9}$$

where h is the height of the frame and EI is the flexural stiffness of the columns, which was defined in the sixth row of Table 3.1. Silva and Badie (2008) gave the following formula for a general case with variable beam-to-column stiffness ratio of $\alpha = I_b/I_c$ (assuming variable moments of inertia, but constant modulus of elasticity) and variable beam-span-to-column-height ratio of $\kappa = L_b/h$:

$$K = \frac{24EI}{h^3}\left(\frac{6\alpha + \kappa}{6\alpha + 4\kappa}\right) \tag{3.10}$$

For a portal frame case with pinned supports and a very stiff beam, K is given by the contribution of two pinned–fixed columns (such as the one listed in the fifth row of Table 3.1) as

$$K = (2)\frac{3EI}{h^3} = \frac{6EI}{h^3} \tag{3.11}$$

Silva and Badie (2008) also gave a general case with variable beam-to-column stiffness ratio of α and variable beam-span-to-column-height ratio of κ:

$$K = \frac{24EI}{h^3}\left(\frac{\alpha}{4\alpha + 2\kappa}\right) \tag{3.12}$$

EXAMPLE 3.7

The bridge systems depicted below has a rigid deck that weighs 500 kips. The lateral force-resisting system consists of three concrete ($E = 3000$ ksi) columns as shown in Figure E3.10. Determine the natural period of the system.

SOLUTION

1. Idealize the structural system into a SDOF system.
   The frame can be modeled as an SDOF system, and the total stiffness of the SDOF system is the sum of the individual column stiffnesses.

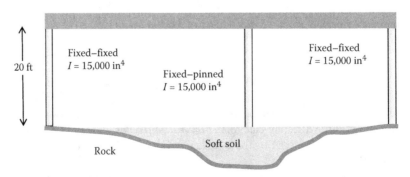

FIGURE E3.10   Schematic of the bridge system.

2. Determine the mass of the SDOF system

$$m = W/g = 500 \text{ kips}/(386.4 \text{ in/s}^2) = 1.294 \text{ kips} \cdot \text{s}^2/\text{in}$$

3. Determine the stiffness.
   The lateral stiffness of the columns depends on the boundary conditions and connections to the bridge deck. In this case, the end conditions of the columns are specified in the problem statement, and the stiffness parameters for each column can be obtained from Table 3.1:

$$K = \frac{12EI}{h^3} + \frac{3EI}{h^3} + \frac{12EI}{h^3} = \frac{27EI}{h^3} = \frac{27(3000)(15,000)}{(20 \cdot 12)^3} = 87.9 \text{ kips/in}$$

4. Determine the natural period of the SDOF system.
   The natural period of the bridge is determined using Equation 3.5:

$$T_n = 2\pi\sqrt{\frac{m}{K}} = 2\pi\sqrt{\frac{1.294}{87.9}} = 0.762 \text{ s}$$

Braced frames can be assumed to behave as cantilever beams that carry the lateral load by developing an internal shear force (known as panel shear) and a bending moment (known as overturning moment). For portal braced frames similar to the one shown in Figure 3.7, we can assume that the panel shear is carried by the diagonal members (one axial compression and the other axial tension), whereas the overturning moment is carried by the axial forces in the columns. This braced frame can be modeled as a statically indeterminate truss in which case the number of unknown reactions, r, and unknown member forces, b, exceeds the number of equations of equilibrium (two times the number of joints), $2j$. To solve for the additional unknown in Figure 3.7, we can use compatibility of displacements. However, because the member sizes are initially unknown, we cannot apply

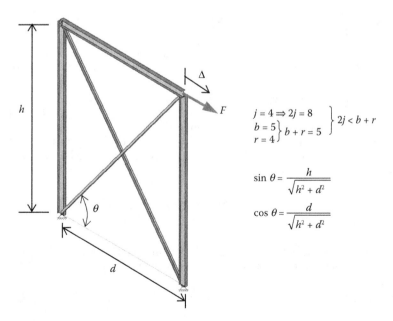

$$j = 4 \Rightarrow 2j = 8$$
$$\left.\begin{matrix} b = 5 \\ r = 4 \end{matrix}\right\} b + r = 5 \left.\vphantom{\begin{matrix} b = 5 \\ r = 4 \end{matrix}}\right\} 2j < b + r$$

$$\sin\theta = \frac{h}{\sqrt{h^2 + d^2}}$$

$$\cos\theta = \frac{d}{\sqrt{h^2 + d^2}}$$

FIGURE 3.7 Braced portal frame model.

displacement compatibility. Therefore, we must make some simplifying assumptions to render the truss statically determinate and perform an approximate analysis. First, the diagonal members can be assumed to be stiff (in which case the diagonals can carry both tension and compression) or they can be assumed to be slender (in which case the member in axial compression cannot support axial force because it buckles). In the case of stiff diagonal members, we assume that the tension and compression diagonals each carry half of the panel shear, whereas in the case of slender diagonals, we assume that the panel shear is resisted entirely by the tension diagonal. The assumption of slender diagonal members is used to derive the equation for the stiffness of a braced portal frame using the principle of virtual work in Example 3.8.

The principle of virtual forces can be used to establish the displacement of a truss joint in any direction. The process entails determining the internal forces in each member caused by the applied loads and determining the virtual internal forces caused by a unit virtual load applied at the joint in the direction in question. With these internal loads, the contribution from each member to the truss deflection is summed to determine the total displacement,

$$\Delta = \sum_{i=1}^{m} n_i \frac{N_i L_i}{E_i A_i} \tag{3.13}$$

where:

N_i is the axial force in each member caused by the actual loading

n_i is the axial force in each member caused by a unit virtual load applied at the joint in the direction in question

L_i is the length of each member

A_i is the area of each member

E_i is the modulus of elasticity of each member

EXAMPLE 3.8

Derive an equation for the stiffness for a braced frame (see Figure 3.7) when the diagonal members are slender.

SOLUTION

1. Apply equilibrium to determine internal forces for actual and virtual loading.

    Slender diagonals cannot carry compression load; thus, from statics we can deduce that the left-hand column and beam are zero-force members, leaving us with the cases shown in Figure E3.11 for both the actual and virtual loading.

    Use the method of joints to determine the internal forces in the remaining members by first drawing an FBD of the upper right-hand joints and applying equilibrium to the resulting concurrent force systems as shown in Figure E3.12.

2. Apply virtual work (Equation 3.13) to determine the displacement at the top of the frame, Δ. As the axial displacement of the column contributes little to the lateral displacement of the frame, it's customary to ignore it:

$$\Delta = \sum n \frac{NL}{EA} = \frac{(1/\cos\theta)(F/\cos\theta)\sqrt{h^2+d^2}}{E_1 A_1} = F\left[\frac{(h^2+d^2)^{3/2}}{d^2 E_1 A_1}\right]$$

3. Determine the stiffness (ratio of force to displacement) of the frame. The frame lateral deflection equation can be rearranged to determine the stiffness as

$$k = \frac{d^2 E_1 A_1}{(h^2+d^2)^{3/2}} \tag{3.14}$$

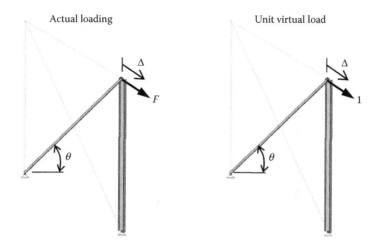

FIGURE E3.11 Actual and virtual loading with slender diagonal elements assumed.

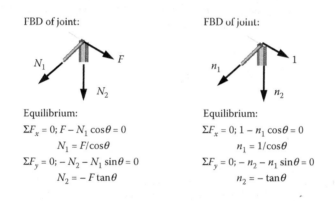

FBD of joint:

Equilibrium:
$\Sigma F_x = 0; F - N_1 \cos\theta = 0$
$\qquad N_1 = F/\cos\theta$
$\Sigma F_y = 0; -N_2 - N_1 \sin\theta = 0$
$\qquad N_2 = -F\tan\theta$

FBD of joint:

Equilibrium:
$\Sigma F_x = 0; 1 - n_1 \cos\theta = 0$
$\qquad n_1 = 1/\cos\theta$
$\Sigma F_y = 0; -n_2 - n_1 \sin\theta = 0$
$\qquad n_2 = -\tan\theta$

FIGURE E3.12 FBDs of joints due to the actual loading (left) and virtual loading (right).

where the quantity EA is the axial stiffness given by the product of the modulus of elasticity, E, and the cross-sectional area, A.

EXAMPLE 3.9

The platform shown in Figure E3.13 is used at a stadium to film football games and is experiencing large dynamic motions. Preliminary investigations indicate that the natural period is 0.9 s. Camera personnel recommend the period to be limited to 0.3 s. Using steel ($E = 29,000$ ksi), determine the required diameter of a system of diagonal ties (wires) to retrofit the system to the new specifications.

SOLUTION

The system can be idealized as an SDOF system. Also, as the platform is square, only one direction of motion needs to be considered, which is retrofitted with two diagonal ties (one along each side—dash lines in Figure E3.13).

1. Determine the existing stiffness of the system using the observed natural period. In terms of mass,

$$m = W/g = 4.5 \text{ kips}/(386.4 \text{ in/s}^2) = 0.01164 \text{ kips} \cdot \text{s}^2/\text{in}$$

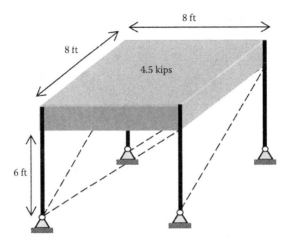

FIGURE E3.13   Schematic of a structural platform.

and the existing natural period of the SDOF system ($T_n = 0.9$ s), we can determine the available stiffness using Equation 3.5:

$$T_n = 2\pi\sqrt{\frac{m}{k}} \Rightarrow k = m\left(\frac{2\pi}{T_n}\right)^2$$

That is,

$$k = m\left(\frac{2\pi}{T_n}\right)^2 = 0.01164\left(\frac{2\pi}{0.9}\right)^2 = 0.5676 \text{ kips/in}$$

2. Determine the required stiffness necessary to correspond to the limiting period. We now determine the required stiffness to deduce the natural period to 0.3 s.

$$k_{\text{new}} = m\left(\frac{2\pi}{T_n}\right)^2 = 0.01164\left(\frac{2\pi}{0.3}\right)^2 = 5.108 \text{ kips/in}$$

3. Determine the required diameter of diagonal ties needed to increase the stiffness of the system. Determine the difference between the available and the new required stiffnesses. This must be provided by the remedial diagonal ties. The change in stiffness can be applied to Equation 3.14 to determine the required radius of the diagonal ties:

$$k_{\text{new}} - k = 2\frac{d^2 EA}{(h^2 + d^2)^{3/2}}$$

where $d = 8$ ft, $h = 6$ ft, and $A = \pi r^2$, so

$$5.108 \times 0.5676 = 2\frac{(8 \times 12)^2(29,000)\pi r^2}{((6 \times 12)^2 + (8 \times 12)^2)^{3/2}}$$

Solving for the radius of the wire yields

$$r = \sqrt{\frac{(5.108 - 0.5676)((6 \times 12)^2 + (8 \times 12)^2)^{3/2}}{2(8 \times 12)^2(29,000)\pi}} = 0.0685 \text{ in}$$

Thus, the required diameter is $2r = 0.137$ in, which can be provided by a 3/16-inch wire.

Shear walls can be assumed to behave as deep cantilever beams that carry the lateral load by developing internal shear force and bending moment as shown in Figure E3.14. That is, a shear wall has deflections that are affected by both moment and shear (unlike regular beams, where most of the deflection is caused by flexure). We can once again use the principle of virtual work to establish the lateral displacement of the top of a shear wall. The process entails determining the internal shear and moment functions (diagrams) caused by the applied lateral load and determining the virtual internal shear force and moment functions (diagrams) caused by a lateral unit virtual load also applied at the top of the wall. With the internal shear force and moment functions, the deflection of the top of the wall can be determined using the following virtual work equation:

$$\Delta = \int_0^L m_f \frac{M_f}{EI} dx + \int_0^L \kappa \left(\frac{vV}{GA} \right) dx \qquad (3.15)$$

where:
m_f is the internal virtual moment caused by an external virtual unit load
M_f is the internal moment caused by the lateral load
v is the internal shear force caused by an external virtual unit load
V is the internal shear force caused by the lateral load
L is the length of the member
I is the moment of inertial of the member
E is the modulus of elasticity of the member
A is the area of the member
G is the shear modulus of elasticity of the member
κ is a constant that takes the following values depending on the cross-section of the shear wall:
  $\kappa = 1.2$ for rectangular beams
  $\kappa = 10/9$ for circular sections
  $\kappa = A/A_{\text{web}}$ (ratio of total area to area of web) for I or box sections

EXAMPLE 3.10

Derive an equation for the stiffness for the cantilever shear wall shown in Figure E3.14.

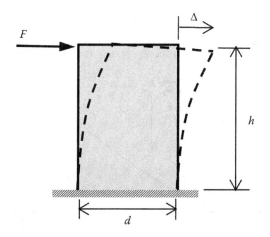

FIGURE E3.14 Cantilever shear wall.

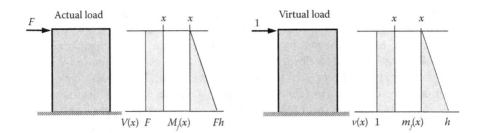

FIGURE E3.15 Internal shear and bending moment diagrams for actual loading and virtual loading.

SOLUTION

1. Apply equilibrium to determine shear force and bending moment functions for actual and virtual loading.
   Use structural analysis to determine the shear force and bending moment diagrams for a cantilever beam, for both real load and virtual loading; see Figure E3.15.
2. Determine the displacement, Δ, at the top of the wall.
   Apply virtual work (Equation 3.15) to determine the displacement at the top of the wall, Δ. As the moments vary linearly and the shears are constant, we get the following result:

$$\Delta = \frac{1}{EI} \int_0^L m_f(x) M_f(x)\, dx + \frac{\kappa}{GA} \int_0^L v(x) V(x)\, dx = \frac{1}{EI3}(h)(Fh)h + \frac{\kappa}{GA}(1)(F)h$$

$$= \frac{Fh^3}{3EI} + \frac{\kappa Fh}{GA}$$

3. Determine the stiffness (ratio of force to displacement) of the wall.
   The wall lateral deflection equation can be rearranged to determine the stiffness, which for a rectangular ($\kappa = 1.2$) shear wall is

$$K = \frac{3EI}{h^3} + \frac{AG}{1.2h} \tag{3.16}$$

where the quantity EI is the flexural stiffness given by the product of the modulus of elasticity, E, and the moment of inertia of the area, I, and AG is the shear stiffness given by the product of the shear modulus of elasticity, G, and the area, A.

Figure 3.8 illustrates the relationship between the shear wall aspect ratio, d/h, and relative effect of shear displacement, Δ/Δ_M. Δ_M is the displacement cause by moment only. Note that as the shear wall gets taller, the effect of the shear force decreases rather quickly. For $d/h < 1/5$, the shear

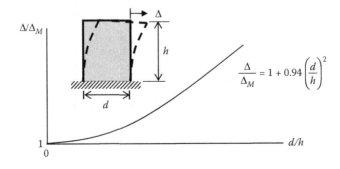

FIGURE 3.8 Effect of shear force in the lateral deflection of a shear wall.

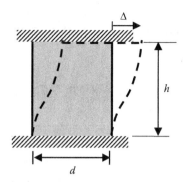

FIGURE 3.9 Fixed top and bottom shear wall.

effect represents <4% so it may be neglected; however, for $d/h > 1/3$, the shear effect is important as it is >10%.

We can also derive the equation for a shear wall that is fixed at the top and bottom (see Figure 3.9),

$$K = \frac{12EI}{h^3} + \frac{AG}{1.2h} \tag{3.17}$$

EXAMPLE 3.11

The structural system depicted in Example 3.6 (repeated here as Figure E3.16 for convenience) has a roof weighing 20 psf and side sheeting weighing 10 psf. Assuming the lateral force-resisting system in the N–S direction (one on each side) consists of ½-inch plywood ($E = 1000$ ksi and $G = 500$ ksi) shear walls, determine the building's natural period in this direction.

SOLUTION

1. Determine the weight (proportional to the mass) of the SDOF system (same as Example 3.6).

    The weight of the SDOF system includes the roof dead load of 20 psf and the wall sheathing weight around the perimeter of the building (240 ft). The weight due to sheathing includes the tributary height of the roof, 10 ft, in the area calculation:

$$W = 20 \text{ psf } (40' \times 80') + 10 \text{ psf } (10' \times 240') = 88,000 \text{ lb} = 88 \text{ kips}$$

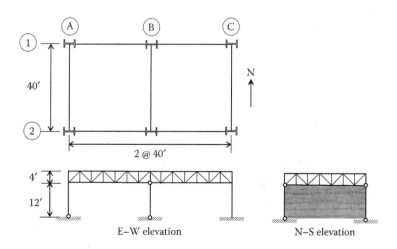

FIGURE E3.16 Plan and elevation views of a structural system.

2. Determine the stiffness parameters.

The lateral stiffness of the shear walls depends on the moment of inertia and area of the footprint of the wall.

Moment of inertia of shear wall,

$$I = bh^3/12 = 0.5 \, \text{in}(20 \, \text{ft} \times 12 \, \text{in/ft})^3/12 \; = \; 576,000 \, \text{in}^4$$

Area of shear wall,

$$A = bh = 0.5 \, \text{in} \, (20 \, \text{ft} \times 12 \, \text{in/ft}) = 120 \, \text{in}^2$$

The total stiffness of the SDOF system is given by Equation 3.16:

$$K = 2\left[\frac{3EI}{h^3} + \frac{AG}{1.2\,h}\right] = 2\left[\frac{3(1000 \, \text{ksi})(576,000 \, \text{in}^4)}{(14 \cdot 12)^3} + \frac{120 \, \text{in}^2(500 \, \text{ksi})}{1.2(14 \cdot 12)}\right] = 1324 \, \text{k/in}$$

3. Determine the natural period of vibration.

As the frame is modeled as an SDOF system, the equation of motion is described by Equation 3.1. The natural period of the system is determined using Equation 3.5:

$$T_n = 2\pi\sqrt{\frac{m}{k}} = 2\pi\sqrt{\frac{W/g}{k}} = 2\pi\sqrt{\frac{88 \, \text{kips}/386.4 \, \text{in/s}^2}{1324 \, \text{kips/in}}} = 0.083 \, \text{s}$$

3.1.4 Approximate Structural Period

For the *equivalent static method* of analysis, the ASCE-7 specifications provide approximate relations to determine the period of a structural system. The following relationships are used to determine an approximate period, T_a:

$$T_a = C_t h_n{}^x \tag{3.18}$$

where:

h_n is the height in feet above the base to the highest level of the structure
C_t and x are determined from ASCE-7 (Table 12.8.2; see Table 3.2)

An even more approximate relation that can be used to quickly check the values determined using Equation 3.18 or the full dynamic analysis period is

$$T_a = 0.1N \tag{3.19}$$

TABLE 3.2

Values of Approximate Period Parameters C_t and x (ASCE-7 Table 12.8.2)

Structure Type	C_t	x
Steel moment-resisting frames	0.028	0.8
Concrete moment-resisting frames	0.016	0.9
Eccentrically braced steel frames	0.03	0.75
All other structural systems	0.02	0.75

---

where N is the number of stories.

This approximate period can be considered valid for buildings of up to 12 stories and story heights uniformly distributed at about 10 ft.

Approximate period for shear walls:

$$T_a = \frac{0.0019}{\sqrt{C_w}} h_n \tag{3.20}$$

where:

$$C_W = \frac{100}{A_B} \sum_{i=1}^{x} \left(\frac{h_n}{h_i}\right)^2 \frac{A_i}{[1+0.83(h_i/D_i)^2]}$$

A_B is the area of the base of the structure in square feet
A_i is the web area of shear wall i in square feet
D_i is the length of shear wall i in feet
h_i is the height of shear wall i in feet

3.1.5 TIME-DEPENDENT FORCED UNDAMPED VIBRATION RESPONSE

We now characterize the forced vibration motion of the portal frame model introduced in Section 3.1.1 (see Figure 3.10). In this case, the mathematical model includes a time-dependent excitation force, $p(t)$ (or displacement). We can again apply equilibrium to an FBD of the mass (which now includes the time-dependent force as shown in Figure 3.10) using D'Alembert's principle. Horizontal equilibrium of the FBD shown in Figure 3.10 yields the equation of motion,

$$\sum F_x = 0; \quad -m\ddot{u} - ku + p(t) = 0 \Rightarrow m\ddot{u} + ku = p(t) \tag{3.21}$$

For the purpose of this analysis, we use a periodic, harmonic excitation force, which is a force with magnitudes represented by sine or cosine as functions of time. (This force can also represented a time-dependent displacement excitation, such as one produced by rotating machinery.) Sine and cosine functions can also be applied to nonharmonic loadings with the response being obtained using a Fourier method—a superposition of individual responses to the harmonic components of the external excitations. First, assume a forcing function of the following form:

$$p(t) = p_o \sin \omega t \tag{3.22}$$

where:

p_o is the peak magnitude of the force
ω is the frequency of the force in radians per second

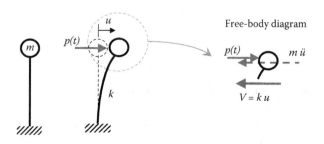

FIGURE 3.10 Free-body diagram of the SDOF model including a time-dependent force.

This forcing function produces bound displacements provided that $\omega \neq \omega_n = \sqrt{k/m}$, the resonant condition, which will be discussed in more detail later in this section.

With a forcing function, we can solve the equation of motion following standard methods used in the solution of nonhomogeneous differential equations. Equation 3.21 can be rewritten as

$$m\ddot{u} + ku = p_o \sin \omega t$$

The general solution to this equation can be expressed as a combination of the particular and complementary solutions,

$$u(t) = u_c(t) + u_P(t) \tag{3.23}$$

where:
  u_c is the complementary solution, which is the solution to the homogeneous equation (the free vibration solution) given by Equation 3.3 and repeated here for convenience,

$$u_c(t) = A \cos \omega_n t + B \sin \omega_n t$$

u_P is the particular solution, which usually takes the same form as the forcing function, that is,

$$u_p(t) = C \sin \omega t \tag{3.24}$$

where C is the peak value, which is obtained by substituting Equation 3.24 into the original equation of motion,

$$-m\omega^2 C \sin \omega t + kC \sin \omega t = p_o \sin \omega t$$

After factoring the sine function that appears in each term, and recognizing that this function is not always zero, we can write this equation as

$$-m\omega^2 C + kC = p_o$$

Solving for C,

$$C = \frac{p_o}{k - m\omega^2} = \frac{p_o}{k(1 - \sqrt{m/k\omega^2})} = \frac{(u_{st})_o}{1 - r^2}$$

where:
  r is the frequency ratio, that is,

$$r = \frac{\omega}{\omega_n}$$

$(u_{st})_o$ is the static displacement caused by p_o; that is, $(u_{st})_o = p_o/k$

We can rewrite the particular solution to include the phase angle, ϕ, as follows:

$$u_P(t) = u_o \sin(\omega t - \phi)$$

where:

u_o is the maximum steady-state displacement response caused by the time-dependent excitation force;

or

$$u_P(t) = (u_{st})_o R_d \sin(\omega t - \phi) \tag{3.25}$$

where:

R_d is the deformation response factor, which is the ratio of the maximum steady-state displacement, u_o, and the equivalent static displacement, $(u_{st})_o$,

$$R_d = \frac{u_o}{(u_{st})_o} = \frac{1}{1 - r^2} \tag{3.26}$$

and the phase angle, ϕ, ranges as follows:

$$\phi = \begin{cases} 0° & \omega < \omega_n \\ 180° & \omega > \omega_n \end{cases}$$

Figure 3.11 shows graphs depicting the deformation response factor, R_d, and phase angle, ϕ, as functions of frequency ratio, r. First, note that the displacement grows unbounded as r approaches one, both from the right and the left. In actual practice, however, the system would come apart or yield (changing the stiffness, thus changing the natural frequency). This point on the graph corresponds to the resonant frequency, and as indicated earlier, the solution to the equation of motion is not valid here. Also, the excitation force changes from being in phase with the natural

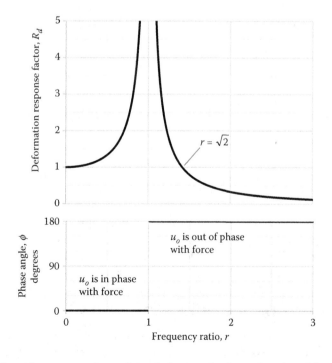

FIGURE 3.11 Deformation response factor, R_d, and phase angle, ϕ, as a function of frequency ratio, r.

vibration of the system to being out of phase; that is, the forcing function is additive up to $r = 1$; and it begins to counter the force after $r = 1$. Furthermore, note that for small values of r (slowly varying force), R_d is ~ 1; and for large values of r (rapidly varying force), R_d approaches zero, which corresponds to no displacement. An additional point of interest along the abscissa of these graphs is $r = \sqrt{2}$, beyond which all values of R_d are ≤ 1. The point can be obtained by setting $R_d = 1.0$. This implies that the dynamic displacement is always less than the static displacement beyond $r = \sqrt{2}$.

We now obtain the total dynamic response of the system by combining the complementary and particular solutions,

$$u(t) = A\cos\omega_n t + B\sin\omega_n t + \frac{p_o/k}{1 - r^2}\sin\omega t \qquad (3.27)$$

The two constants A and B are arbitrary and can be obtained by evaluating the equation at time $t = 0$ (initial conditions). Again, the displacement at time $t = 0$ is $u(0)$, that is,

$$u(0) = A\cos(0) + B\sin(0) + \frac{p_o/k}{1 - r^2}\sin(0)$$

which again yields $A = u(0)$. Next, differentiate the displacement equation with respect to time,

$$\dot{u}(t) = -A\omega_n \sin\omega_n t + B\omega_n \cos\omega_n t + \omega\frac{p_o/k}{1 - r^2}\cos\omega t \qquad (3.28)$$

Evaluating the velocity at time $t = 0$,

$$\dot{u}(0) = -A\omega_n \sin(0) + B\omega_n \cos(0) + \omega\frac{p_o/k}{1 - r^2}\cos(0)$$

yields

$$B = \frac{\dot{u}(0)}{\omega_n} - \frac{p_o/k}{1 - r^2}r$$

The complete solution is the forced vibration of an undamped SDOF system, which describes the position of the mass as a function of time,

$$u(t) = u(0)\cos\omega_n t + \left(\frac{\dot{u}(0)}{\omega_n} - \frac{p_o/k}{1 - r^2}r\right)\sin\omega_n t + \frac{p_o/k}{1 - r^2}\sin\omega t \qquad (3.29)$$

A graph of which is given in Figure 3.12. In general, the resulting motion is the sum of periodic motions of two different frequencies, ω and ω_n, with different amplitudes. As noted earlier, when ω approaches ω_n, the motion becomes unbounded as t goes to infinity. However, when damping (discussed in Section 3.2) is included, the motion remains bounded.

3.2 DAMPED SINGLE-DEGREE-OF-FREEDOM SYSTEM

An undamped system, once excited, will oscillate indefinitely with constant amplitude at its natural frequency; however, from experience we know that no systems can oscillate indefinitely. Any system undergoing motion always experiences internal damping forces that dissipate mechanical

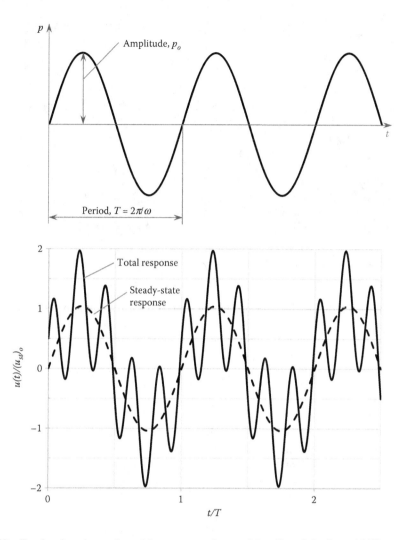

FIGURE 3.12   Forcing function and resulting response for $r = 0.2$, $u(0) = 0.5p_o/k$, and $\dot{u}(0) = \omega_n p_o/k$.

energy (kinetic and potential) by transforming it into other forms of energy, such as heat. This type of damping is inherent to a system. There may be cases where intentional damping is introduced in a system, such as the shock absorbers in an automobile. Similar types of shock absorbers can also be used for building systems, such as the fluid viscous dampers from Taylor Devices shown in Figure 3.13. Similar to inertial and stiffness effects, damping effects can be characterized using a force, which in this case is assumed to be proportional to the magnitude of the velocity and opposite to the direction of motion. This type of damping is known as "viscous damping" and is the type of damping produced in a body when restrained by surrounding viscous fluid. While viscous damping is not inherent in structural systems, this type of damping is much easier to manipulate in the equation of motion than the more realistic damping from internal friction and is accurate enough for most practical purposes.

3.2.1 Free Vibration Response of Damped Systems

We can include the damping effects in the free vibration response of the portal frame model discussed in Section 3.1.1 by first applying equilibrium to an FBD of the mass using D'Alembert's

FIGURE 3.13   Fluid viscous dampers (photographs courtesy of Taylor Devices, Inc.).

principle, where, in addition to the stiffness and inertial forces, a damping force proportional to the velocity is included, as shown in Figure 3.14. In this case, we are using an oscillator model, which is commonly used in structural dynamics textbooks (Chopra, 2012) because it is easier to visualize the combined effects of damping, stiffness, and inertia.

Horizontal equilibrium of the FBD shown in Figure 3.14 yields the equation of motion,

$$m\ddot{u} + c\dot{u} + ku = 0 \tag{3.30}$$

Here, c is the damping coefficient, which is discussed in more detail in the next section.

Again, this is a second-order, linear, and homogeneous differential equation with constant coefficients, the solution to which is also of exponential form, that is,

$$u(t) = e^{\rho t}$$

This equation can be differentiated once to determine the velocity,

$$\dot{u}(t) = \rho e^{\rho t}$$

and the velocity can be differentiated to determine the acceleration,

$$\ddot{u}(t) = \rho^2 e^{\rho t}$$

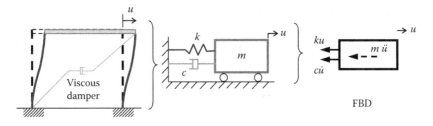

FIGURE 3.14   Idealized SDOF system and FBD for a portal frame.

substituting the equations for the displacement, velocity, and the acceleration into Equation 3.30 yields

$$m\rho^2 e^{\rho t} + c\rho e^{\rho t} + k e^{\rho t} = 0$$

which can be rearranged as follows:

$$(m\rho^2 + c\rho + k)e^{\rho t} = 0$$

In general, the exponential function is not zero, thus the quantity within parenthesis (also known as the characteristic equation) must be zero,

$$m\rho^2 + c\rho + k = 0$$

Solving for the unknown parameter ρ yields

$$\rho = \frac{-c \pm \sqrt{c^2 - 4mk}}{2m} \Rightarrow \begin{cases} \rho_1 = \dfrac{-c + \sqrt{c^2 - 4mk}}{2m} \\[2ex] \rho_2 = \dfrac{-c - \sqrt{c^2 - 4mk}}{2m} \end{cases}$$

Since the equation of motion (Equation 3.30) is a second-order differential equation, we need two constants of integration, C_1 and C_2, which are determined from initial conditions (displacement and velocity).

$$u(t) = C_1 e^{\rho_1 t} + C_2 e^{\rho_2 t}$$

Substituting the ρ parameters into this equation yields

$$u(t) = e^{\frac{-c}{2m}t}\left(C_1 e^{\frac{\sqrt{c^2 - 4mk}}{2m}t} + C_2 e^{\frac{-\sqrt{c^2 - 4mk}}{2m}t} \right)$$

This result leads to three different solutions depending on the value of the quantity under the square root operator (see Figure 3.15):

1. If $c^2 - 4mk > 0$, exponents are real numbers. The motion for this case decays exponentially; thus, no oscillation occurs. For this case, the system is classified as *overdamped*.
2. If $c^2 - 4mk < 0$, exponents are imaginary values, and oscillatory motion occurs. For this case, the system is classified as *underdamped*.
3. If $c^2 - 4mk = 0$, then there is a boundary between oscillatory and nonoscillatory motion. For this case, the system is classified as *critically damped*.

Structural systems are typically underdamped, so our discussion on damping will be focused primarily on this type of system. However, critically damped systems are important because they are used to define the damping ratio, ζ, which is typically used to describe damping in structural systems. For a critically damped system, $c^2 - 4mk = 0$, which leads to the definition of the critical damping coefficient, c_{cr}:

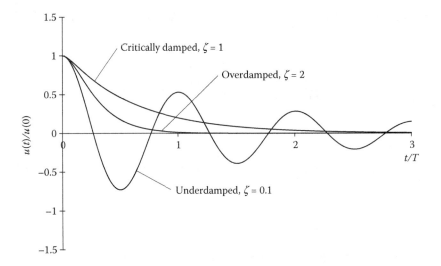

FIGURE 3.15   Free vibration response of a damped SDOF system.

$$c_{cr} = 2\sqrt{km} = 2m\omega_n = \frac{2k}{\omega_n} \tag{3.31}$$

The other parameters were defined earlier. With the critical damping coefficient, c_{cr}, we now define the damping ratio or damping factor (which is an expedient way to define damping for practical structural systems and is usually expressed in a percentage form) as

$$\zeta = \frac{c}{c_{cr}} \tag{3.32}$$

We can also use this value to define the three different general states of damping: overdamped systems have $\zeta > 1$, critically damped systems have $\zeta = 1$, and underdamped systems have $\zeta < 1$, as shown in Figure 3.15. Since most structural systems are underdamped, in this book we only provide the solution for this case. Rewriting the unknown parameters ρ_1 and ρ_2 as

$$\rho_1, \rho_2 = \left(-\zeta \pm \sqrt{\zeta^2 - 1}\right)\omega_n$$

This will be complex as $\zeta < 1$. We can write this in complex number form as

$$\rho_1, \rho_2 = \left(-\zeta \pm i\sqrt{1 - \zeta^2}\right)\omega_n$$

Substituting these parameters into the solution yields

$$u(t) = e^{-\zeta\omega_n t}\left[C_1 e^{i\sqrt{1-\zeta^2}\,\omega_n t} + C_2 e^{-i\sqrt{1-\zeta^2}\,\omega_n t}\right]$$

This equation can be expressed in polar form (in terms of sines and cosines) by making use of Euler's identities; and because trigonometric functions are real-value solutions to the equation of motion, we can express the general solution in real form as

$$u(t) = e^{-\zeta\omega_n t}[B\sin\omega_D t + A\cos\omega_D t] \tag{3.33}$$

where A and B are arbitrary constants. While this motion is periodic, it is no longer of constant frequency. In fact, the frequency (and the period) is a function of the damping ratio—the damped frequency,

$$\omega_D = \omega_n\sqrt{1-\zeta^2} \tag{3.34}$$

The damped period,

$$T_D = \frac{2\pi}{\omega_D} = \frac{2\pi}{\omega_n\sqrt{1-\zeta^2}} = \frac{T_n}{\sqrt{1-\zeta^2}} \tag{3.35}$$

We now use initial conditions (at time $t = 0$) to solve for constants A and B. First, the initial displacement (at time $t = 0$) is $u(0)$, that is,

$$u(0) = e^0[B\sin(0) + A\cos(0)]$$

which yields $A = u(0)$. Next, differentiating the displacement equation with respect to time we obtain

$$\dot{u}(t) = -\zeta\omega_n e^{-\zeta\omega_n t}[B\sin\omega_D t + A\cos\omega_D t] + e^{-\zeta\omega_n t}[B\omega_D\cos\omega_D t + A\omega_D\sin\omega_D t]$$

Evaluating the velocity at time $t = 0$ yields

$$\dot{u}(0) = -\zeta\omega_n e^0[B\sin(0) + A\cos(0)] + e^0[B\omega_D\cos(0) + A\omega_D\sin(0)]$$

where:

$$B = \frac{\dot{u}(0) + \zeta\omega_n u(0)}{\omega_D}$$

Again, the complete solution is the free vibration response of a damped SDOF system, which describes the position of the mass as a function of time,

$$u(t) = e^{-\zeta\omega_n t}\left[\left(\frac{\dot{u}(0) + \zeta\omega_n u(0)}{\omega_D}\right)\sin\omega_D t + u(0)\cos\omega_D t\right] \tag{3.36}$$

A graph of which is given in Figure 3.16. The damped structure clearly shows motion decay over time. The decay in the motion is mathematically produced by the exponential part of the solution.

This oscillation decay can be used to experimentally determine the inherent damping in a system. The process entails setting the structure into free vibration from an initial displacement and obtaining a record of the motion. The rate of decay of amplitude of motion is characterized by a ratio of the value of two successive peak displacement amplitudes, which is known as the logarithmic decrement, δ. The values of two successive peak displacement amplitudes are

$$u_1 = \iota e^{-\zeta\omega_n t_1} \qquad u_2 = \iota e^{-\zeta\omega_n(t_1 + T_D)}$$

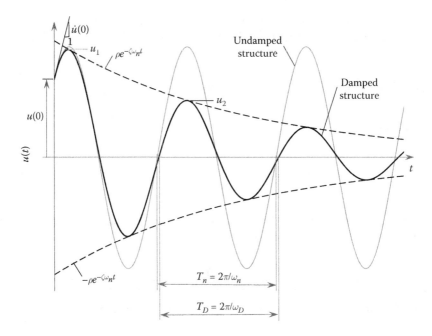

FIGURE 3.16 Comparison of free vibration response of undamped and damped SDOF systems.

where ι is a constant that depends on the initial conditions. The ratio of these displacements is

$$\frac{u_1}{u_2} = \frac{\iota e^{-\zeta \omega_n t_1}}{\iota e^{-\zeta \omega_n (t_1 + T_D)}} = e^{\zeta \omega_n (t_1 + T_D) - \zeta \omega_n t_1} = e^{\zeta \omega_n T_D}$$

Taking the natural log of both sides, we can define the logarithmic decrement, δ,

$$\delta = \ln\left(\frac{u_1}{u_2}\right) = \zeta \omega_n T_D \tag{3.37}$$

The logarithmic decrement can also be defined in terms of acceleration as

$$\delta = \ln\frac{\ddot{u}_1}{\ddot{u}_2} = \zeta \omega_n T_D \tag{3.38}$$

In both cases, δ can be defined in terms of ζ only. This is accomplished by replacing the damped period as follows:

$$\delta = \zeta \omega T_D = \frac{\zeta \omega_n 2\pi}{\omega_n \sqrt{1 - \zeta^2}} = \frac{2\pi \zeta}{\sqrt{1 - \zeta^2}} \tag{3.39}$$

For small damping ratio values ($\zeta < 0.2$), the value under the square root operator is close to unity and the logarithmic decrement can be approximated as

$$\delta \approx 2\pi\zeta \tag{3.40}$$

Furthermore, for lightly damped systems where successive oscillation peaks have similar ordinates, two nonconsecutive amplitudes u_i and u_{i+n} (where n is any integer) can be used to determine δ,

$$\ln \frac{u_i}{u_{i+n}} = n\zeta\omega_n T_D = n\delta \tag{3.41}$$

Also, note that the equation of motion can be rewritten in the following form:

$$\ddot{u} + 2\zeta\omega_n\dot{u} + \omega_n^2 u = 0 \tag{3.42}$$

3.2.2 Structural Damping

As noted earlier, in addition to attenuating vibration amplitudes, damping has the effect of lowering the natural frequency, and thus lengthening the period from the natural state. Most structural systems possess small inherent damping, $<20\%$ as listed in Table 3.3; and as shown in Figure 3.17, damping effects on the period are negligible when the damping ratio is low. Thus, the damped and undamped periods are practically equal, and in practice, T_D is usually taken as T_n (except when supplemental damping is added to a structural system).

The damping ratio characterizes how fast oscillations decay from one cycle of motion to the next and has a significant effect on the response of elastic systems. However, the inelastic response is dominated by ductility, and damping has little attenuation effect relative to ductility—this will be described in Chapter 5. For this reason, seismic design guidelines (which assume inelastic behavior) assume a uniform 5% damping for all building structural systems. Furthermore, current knowledge of damping mechanisms in structural systems, particularly buildings, is rather limited, with little ongoing research in the area. It is believed that damping in building systems is the result of the following damping mechanisms: friction in structural elements (friction in bolted and nailed connections, and cracking of concrete), intrinsic material damping, friction in nonstructural components and their connections to the structure, and soil–structure interaction (radiation of seismic waves back into the soil and intrinsic damping in the soil) (Villaverde, 2009). All of the aforementioned damping mechanisms are difficult to characterize, thus making it difficult to determine damping analytically for actual structures. Damping ratios tabulated in Table 3.3 (Lindeburg and

TABLE 3.3
Typical Damping Ratios

Type of Construction	ζ
Steel frame with welded connections and flexible walls	0.02
Steel frame with welded connections, normal floors, and exterior cladding	0.05
Steel frame with bolted connections, normal floors, and exterior cladding	0.10
Concrete frame with flexible interior walls	0.05
Concrete frame with flexible interior walls and exterior cladding	0.07
Concrete frame with concrete or masonry shear walls	0.10
Concrete or masonry shear walls	0.10
Wood frame and shear walls	0.15

Source: Lindeburg, M. R. and K. M. McMullin, *Seismic Design of Building Structures: A Professional Introduction to Earthquake Forces and Design Details*, 9th edition, Professional Publications, Inc., Belmont, CA, 2008.

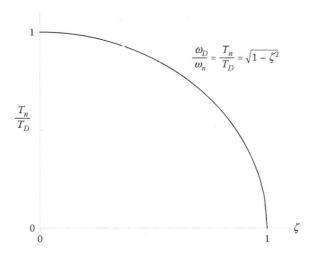

FIGURE 3.17 Effects of damping on vibration period.

McMullin 2008) are only provided to illustrate that real structures do not possess inherent damping >15%. From the given data, it should also be clear that the damping ratio depends on the type of building construction.

EXAMPLE 3.12

A free vibration test is conducted on an empty elevated water tank. A cable attached to the tank applies a lateral force of 16.4 kips and pulls the tank horizontally by 2 in. The cable is suddenly cut and the resulting free vibration is recorded. At the end of four complete cycles, the time is 2.0 s, and the amplitude is 1 in. From the given information, compute the damping ratio, natural period, effective stiffness, effective weight, damping coefficient, and number of cycles required for the displacement to decrease to 0.2 in.

SOLUTION

1. Determine the dampin
   As $u_i = u_1 = 2$ in at $t = 0$ and $u_{i+n} = u_5 = 1$ in at $n = 4$, from Equation 3.41

$$\ln\left(\frac{u_1}{u_5}\right) = \ln\left(\frac{2}{1}\right) = n\zeta 2\pi$$

Therefore, $\zeta = \dfrac{1}{4 \times 2\pi} \ln\left(\dfrac{2}{1}\right) = 0.0276$ or 2.76%.

The low damping ratio allows for a small damping assumption.

2. The damped period is determined by dividing the lapse time by the number of cycles,

$$T_D = \frac{2\,\text{s}}{4\,\text{cycles}} = 0.5\,\text{s/cycle}$$

In addition, due to the small damping assumption, the natural period can be approximated as the damped period.

$$T_n \cong T_D = 0.5\ \text{s}$$

3. Determine the equivalent stiffness of the water tank assuming a linear force–displacement relationship

$$k = \frac{F}{u} = \frac{16.4 \text{ kips}}{2 \text{ in}} = 8.2 \text{ kips/in}$$

4. Determine the effective weight using the period and the stiffness

$$\omega_n = \frac{2\pi}{T_n} = \frac{2\pi}{0.5 \text{ s}} = 12.57 \text{ rad/s}$$

$$m = \frac{k}{\omega_n^2} = \frac{8.2 \text{ kips/in}}{(12.57 \text{ rad/s})^2} = 0.0519 \text{ kip} \cdot \text{s}^2/\text{in}$$

$$W = mg = (0.0519 \text{ kip} \cdot \text{s}^2/\text{in})(386.4 \text{ in/s}^2) = 20.05 \text{ kips}$$

5. The damping coefficient

$$c = \varsigma(2\sqrt{Km}) = 0.0276\left(2\sqrt{(8.2 \text{ kips/in})(0.0519 \text{ kip} \cdot \text{s}^2/\text{in})}\right) = 0.036 \text{ kip} \cdot \text{s/in}$$

6. Determine the number of cycles for displacement to reduce to 0.2 inches using the logarithmic decrement δ

$$\ln\left(\frac{u_i}{u_{i+n}}\right) = n\varsigma 2\pi \Rightarrow$$

$$n = \frac{1}{\varsigma 2\pi}\ln\left(\frac{u_i}{u_{i+n}}\right) = \frac{1}{(0.0276)2\pi}\ln\left(\frac{2 \text{ in}}{0.2 \text{ in}}\right) = 13.3 \text{ cycles} \quad \text{or} \quad 13 \text{ cycles}$$

3.2.3 TIME-DEPENDENT FORCED DAMPED VIBRATION RESPONSE

We now characterize the forced damped vibration motion of the portal frame discussed in Section 3.1.5, shown in Figure 3.10, including the damping force in the mathematical model. The equilibrium equation in this case yields an equation of motion of the form

$$m\ddot{u} + c\dot{u} + ku = p(t) \tag{3.43}$$

For the purpose of this analysis, we again use a periodic, harmonic excitation force. The force excitation can also be represented as a time-dependent displacement excitation as will be shown in Section 3.2.4. We begin again by assuming a forcing function of the form

$$p(t) = p_o \sin \omega t$$

where:
p_o is the peak magnitude of the force
ω is the frequency of the force in radians per second

This forcing function produces bound displacements; even for the resonant condition, provided damping is nonzero. Again, with this forcing function, we can solve the equation of motion,

$m\ddot{u} + c\dot{u} + ku = p_o \sin \omega t$, following standard methods used in the solution of nonhomogeneous differential equations. The general solution to this equation is of the same form as that for the undamped case,

$$u(t) = u_c(t) + u_P(t) \tag{3.44}$$

where u_c is the complementary solution, which is the solution to the homogeneous equation for the underdamped case (the free vibration solution) given by Equation 3.33, and repeated here for convenience,

$$u_c(t) = e^{-\zeta \omega_n t}[A \cos \omega_D t + B \sin \omega_D t]$$

u_p is the particular solution, which usually takes the same form as the forcing function, that is,

$$u_P(t) = C \sin \omega t + D \cos \omega t \tag{3.45}$$

where C and D are constants that are obtained by substituting the particular solution into the original equation of motion,

$$C = \frac{p_o}{k} \frac{(1-r^2)}{(1-r^2)^2 + (2\zeta r)^2} \quad \text{and} \quad D = -\frac{p_o}{k} \frac{2\zeta r}{(1-r^2)^2 + (2\zeta r)^2}$$

Substituting the constants into the particular solution yields

$$u_P(t) = \frac{p_o/k}{(1-r^2)^2 + (2\zeta r)^2}((1-r^2)\sin \omega t - 2\zeta r \cos \omega t) \tag{3.46}$$

We can rewrite the particular solution to include the phase angle, ϕ, as follows:

$$u_P(t) = u_o \sin(\omega t - \phi)$$

where:

$$u_o = \sqrt{C^2 + D^2} = \frac{(u_{st})_o}{\sqrt{(1-r^2)^2 + (2\zeta r)^2}}$$

is the maximum steady-state displacement response and

$$\phi = \tan^{-1}(-D/C) = \tan^{-1}\left(\frac{2\zeta r}{1-r^2}\right)$$

$(u_{st})_o = p_o/k$ is the equivalent static displacement
And as given earlier, the frequency ratio is $r = \omega/\omega_n$.
We can again define the deformation response factor (also known as the dynamic magnification factor), R_d, as the ratio of the maximum steady-state displacement, u_o, to the equivalent static displacement, $(u_{st})_o$:

$$R_d = \frac{u_o}{u_{st}} = \frac{1}{\sqrt{(1-r^2)^2 + (2r\zeta)^2}} \tag{3.47}$$

Note that this ratio is a function of r and ζ only. Maximizing R_d in terms of r yields several interesting conditions. First, take the partial derivative of R_d with respect to r is

$$\frac{\partial R_d}{\partial r} = \frac{r(1-r^2-2\zeta^2)}{[(1-r^2)^2+(2r\zeta)^2]^{3/2}} = 0$$

This leads to four conditions:

1. $r = 0$ leads to $1-2\zeta^2 = 0$, or $\zeta = \sqrt{0.5}$,
2. $r = \infty$ does not yield any useful information,
3. $r = 1$ leads to resonance, $R_{dres} = 1/2\zeta$, so, as $\zeta \to 0$, $R_{dres} \to \infty$,
4. r other than 0, 1, or infinity yields $1-r^2-2\zeta^2 = 0 \Rightarrow r = \sqrt{1-2\zeta^2}$.

Figure 3.18 shows graphs depicting the deformation response factor, R_d, as a function of frequency ratio, r. This graph is the same as that shown in Figure 3.11, but includes damping. Notice that in this case, the displacement does not grow unbounded as r approaches 1. In fact, as damping increases the displacement at the resonant frequency decreases, and for damping ratios exceeding 70.7%, the dynamic displacement is always less than the static displacement. Again, for small values of r (slowly varying force), R_d is approximately equal to 1; and for large values of r (rapidly varying force), R_d approaches zero, which corresponds to no displacement.

We now obtain the total dynamic response of the system by combining the complementary solution (transient response) and particular solution (steady-state response),

$$u(t) = e^{-\zeta\omega_n t}(A\cos\omega_D t + B\sin\omega_D t)$$
$$+ \frac{p_o/k}{(1-r^2)^2+(2\zeta r)^2}((1-r^2)\sin\omega t - 2\zeta r\cos\omega t) \tag{3.48}$$

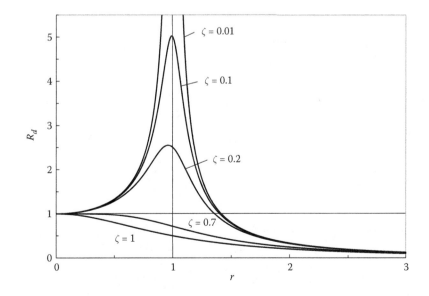

FIGURE 3.18 Damped R_d as a function of r.

The two constants A and B are arbitrary and can be obtained using the initial (at time $t = 0$) conditions. Again, the displacement at time $t = 0$ is $u(0)$, which yields the following when substituted into Equation 3.48:

$$A = u(0) - \frac{(u_{st})_o \sin(-\phi)}{\sqrt{(1-r^2)^2 + (2\zeta r)^2}}$$

Evaluating the velocity at time $t = 0$ yields

$$B = \frac{1}{\omega_D} \left\{ \dot{u}(0) + u(0)\zeta\omega - \frac{(u_{st})_o(\omega\cos(-\phi) + \zeta\omega\sin(-\phi))}{\sqrt{(1-r^2)^2 + (2\zeta r)^2}} \right\}$$

A graph of the complete solution that describes the position of the mass as a function of time is given in Figure 3.19 (similar to Figure 3.12, but with damping). In general, the resulting motion is the sum of periodic motions of two different frequencies (ω and ω_n) and amplitudes. Note that the transient response (the one due to the inherent motion in the system) attenuates as t goes to infinity. Also, as noted earlier for undamped systems, the motion becomes unbounded when ω approaches ω_n, as t goes to infinity. However, for damped cases, motion remains bounded to a maximum of 0.5ζ, as shown in Figure 3.20.

The effect of a harmonic force excitation on the foundation of a supporting system can be determined as the magnitude of the force transmitted to the support through the spring (stiffness) and the damping elements,

$$(f_T)_o = ku_p + c\dot{u}_p = (u_{st})_o R_d[k\sin(\omega t - \phi) + c\omega\cos(\omega t - \phi)]$$

The maximum value of this force over t is

$$(f_T)_o = (u_{st})_o R_d\sqrt{k^2 + c^2\omega^2}$$

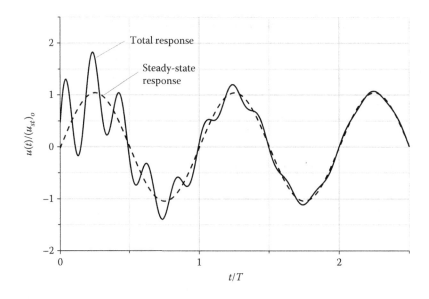

FIGURE 3.19   Harmonic force response for $r = 0.2$, $\zeta = 0.05$, $u(0) = 0$, and $\dot{u}(0) = \omega_n p_o/k$.

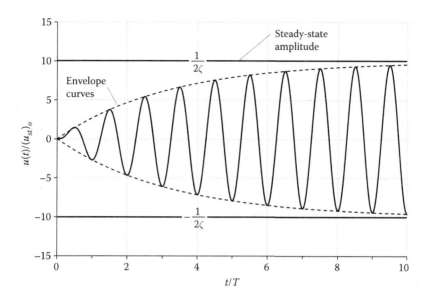

FIGURE 3.20 Resonant response ($r = 1$) to harmonic force for $\zeta = 0.05$.

Rewriting this as a fraction of equivalent static force, p_o, we get the *transmissibility*, T_r,

$$\frac{(f_T)_o}{p_o} = R_d\sqrt{1+(2\zeta r)^2}$$

or

$$T_r = \frac{(f_T)_o}{p_o} = \frac{\sqrt{1+(2r\zeta)^2}}{\sqrt{(1-r^2)^2+(2r\zeta)^2}} \tag{3.49}$$

EXAMPLE 3.13

A sensitive, 50-lb instrument that requires insulation from vibration is being installed in a building where a reciprocating machine is in use. The machine causes the building floor to vibrate in a harmonic motion at frequency of 1000 cycles per minute. The equipment is going to be installed on four springs of equal stiffness and negligible damping. Determine the stiffness of each spring if the amplitude of the transmitted vibration is to be limited to <15% of the floor vibration.

SOLUTION

After idealizing the system as an SDOF system, use transmissibility with $\zeta = 0$ to determine the required stiffness to limit the vibration transmitted.

$$T_r = \frac{\sqrt{1+(2r\zeta)^2}}{\sqrt{(1-r^2)^2+(2r\zeta)^2}} = \frac{1}{\pm(1-r^2)}$$

1. Determine the maximum required frequency ratio, r. Since $T_r < 1, r > \sqrt{2}$, so, apply the negative sign throughout the denominator to obtain

$$T_r = \frac{1}{(r^2-1)}$$

and establish an inequality such that the transmissibility ratio is <0.15, that is,

$$\frac{1}{(r^2 - 1)} < 0.15$$

solving this inequality for r, we determine that frequency ratio, r, must be >2.77.

2. Determine the required natural cyclical frequency. Use the frequency ratio in the following form:

$$r = \frac{f}{f_n} > 2.77$$

where; $f = 1000$ cycles/min (1 min/60 s) = 16.67 Hz.

The frequency ratio equation then gives a natural cyclical frequency of

$$\frac{16.67\,\text{Hz}}{f_n} > 2.77 \Rightarrow f_n < 6.02\,\text{Hz}$$

3. Determine the mass

$$m = W/g = 50\,\text{lbs}/(386.4\,\text{in/s}^2) = 0.1294\,\text{lbs}\cdot\text{s}^2/\text{in}$$

4. We now determine the total required stiffness to limit the vibration transmitted.

$$f_n = \frac{1}{2\pi}\sqrt{\frac{k_t}{m}} < 6.02\,\frac{1}{\text{s}}$$

Solving for the total stiffness, $k_t < 185$ lbs/in.

As the instrument is supported on four springs, the stiffness of each spring must be <46.3 lbs/in in order to limit the transmitted vibration to <15%.

3.2.4 TIME-DEPENDENT SUPPORT ACCELERATIONS

There are a number of cases where the support (or foundation) of a system is subjected to time-varying displacements, such as seismic wave excitations, which is the main topic of this book. The dynamic response of a system to support motion is similar to that of the same system subjected to a time-varying force as discussed in the last section. Let us characterize the excitation motion of the portal frame discussed in Section 3.1.5, shown in Figure 3.21; including the damping force in the mathematical model. Here, we have three different displacements: ground, $u_g(t)$, relative, $u(t)$, and total with respect to the reference axis, $u^t(t)$. This can be accomplished by establishing the equation

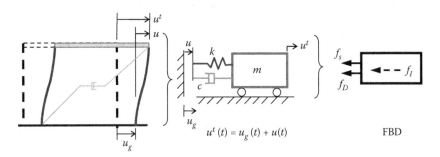

FIGURE 3.21   Idealized SDOF system and free-body diagram for a portal frame.

of motion for the system by applying equilibrium to the FBD of the mass using D'Alembert's principle, where the stiffness and damping forces are proportional to the relative displacement and the inertial force to the total mass displacement. Again, we are using an oscillator model because it is easier to visualize the combined effects of damping, stiffness, and inertia.

Horizontal equilibrium of the FBD shown in Figure 3.21 yields the equation of motion,

$$\sum F_x = 0; \quad f_I(t) + f_D(t) + f_S(t) = 0$$

where:

$f_I(t) = m\ddot{u}^t(t)$ is the inertial force

$f_D(t) = c\dot{u}(t)$ is the damping force and $f_S(t) = ku(t)$ is the stiffness force

So the equation of motion can be rewritten as

$$m\ddot{u}^t(t) + c\dot{u}(t) + ku(t) = 0$$

This equation can be further rewritten in terms of the relative and ground motion by substituting the definition of total acceleration as $\ddot{u}^t(t) = \ddot{u}(t) + \ddot{u}_g(t)$ into the equation of motion,

$$m\ddot{u}(t) + c\dot{u}(t) + ku(t) = -m\ddot{u}_g(t) \tag{3.50}$$

The right-hand side is the effective support excitation loading that opposes the sense of ground acceleration and is similar to the time-varying force discussed in Section 3.1.5.

Alternatively, we can write this equation in terms of the total displacement, in which case the right-hand side is the effective loading; the resulting response is the total displacement of the mass from a fixed reference, rather than displacement relative to the moving base:

$$m\ddot{u}^t + c\dot{u}^t + ku^t = ku_g(t) + c\dot{u}_g(t) \tag{3.51}$$

Before we consider an earthquake excitation, let us analyze a system subjected to a periodic, harmonic base excitation (such as the dynamic action of machinery) given by

$$u_g(t) = u_{go} \sin \omega t$$

where:

u_{go} is the peak amplitude of the ground displacement

ω is the frequency of the support motion in this case

Substituting this displacement relation into Equation 3.51 yields

$$m\ddot{u}^t + c\dot{u}^t + ku^t = ku_{go} \sin \omega t + c\omega u_{go} \cos \omega t \tag{3.52}$$

The two terms on the right-hand side can be combined into an equivalent form by performing a trigonometric transformation such as the one performed in Section 3.1.1 to obtain $u(t) = C \sin(\omega_n t + \alpha)$,

$$m\ddot{u}^t + c\dot{u}^t + ku^t = F_o \sin(\omega t + \alpha) \tag{3.53}$$

where:

$$F_o = \sqrt{(ku_{go})^2 + (c\omega u_{go})^2} = u_{go}\sqrt{k^2 + (c\omega)^2} = u_{go}k\sqrt{1 + (2r\zeta)^2}$$

$$\tan\alpha = \frac{c\omega}{k} = 2r\zeta$$

Equation 3.53 is the differential equation for the oscillator excited by the harmonic force $F_o \sin(\omega t + \alpha)$; the solution to which is of the same form as that for Equation 3.43. That is, following standard methods used in the solution of nonhomogeneous differential equations such as the solution in Equation 3.44, which includes the complementary (transient) and particular (steady state) solutions. The steady-state solution describes the relative transmission of support motion to the mass,

$$u^t(t) = \frac{F_o/k \sin(\omega t + \alpha - \phi)}{\sqrt{(1 - r^2)^2 + (2r\zeta)^2}}$$

or

$$\frac{u^t(t)}{u_{go}} = \frac{\sqrt{1 + (2r\zeta)^2} \sin(\omega t + \alpha - \phi)}{\sqrt{(1 - r^2)^2 + (2r\zeta)^2}} \qquad (3.54)$$

Maximizing this relationship leads to the expression for *transmissibility*, which was discussed earlier and given by Equation 3.49. That is, transmissibility of motion from the foundation to the structure is given by the same function as the transmissibility of force from the structure to the foundation. This is an important expression in vibration isolation; in this context, it is the degree of relative isolation defined as the ratio of the amplitude of motion of the system to the static displacement:

$$T_r = \frac{u_o^t}{u_{st}} = \frac{\sqrt{1 + (2r\zeta)^2}}{\sqrt{(1 - r^2)^2 + (2r\zeta)^2}} = \frac{u_o^t}{u_{go}} \qquad (3.55)$$

where:
$u_{st} = F_o/k$ is the static displacement

Again, we note that this is a function of the frequency ratio, r, and the damping ratio, ζ, only. This can also be expressed in terms of acceleration,

$$T_r = \frac{\ddot{u}_o^t}{\ddot{u}_{go}} = \frac{\sqrt{1 + (2r\zeta)^2}}{\sqrt{(1 - r^2)^2 + (2r\zeta)^2}} \qquad (3.56)$$

Figure 3.22 shows graphs depicting the transmissibility as a function of frequency ratio, r, for various values of ζ. This graph is similar to that shown in Figure 3.18.

From this graph, the following observations are noted:

1. All curves pass through the same point at $r = \sqrt{2}$.
2. Increasing damping when $r < \sqrt{2}$ increases the effectiveness of the vibration isolation system.

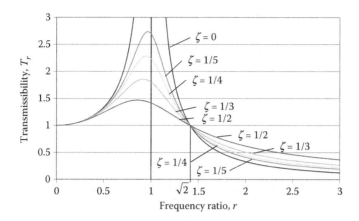

FIGURE 3.22 T_r versus r for various values of ζ.

3. As r approaches zero (slowly varying displacements), $u_o^t = u_{go}$; that is, the mass moves rigidly with the ground motion.
4. $r = 1$ again leads to *resonance*; as $\zeta \to 0$, $T_r \to \infty$.
5. Increasing damping when $r > \sqrt{2}$ decreases the effectiveness of the vibration isolation system.
6. As r approaches infinity (rapidly varying displacements), $u_o^t = 0$; that is, the mass stays still while the ground beneath it moves.

These observations lead to two important conclusions regarding vibration isolation:

1. Since $T_r > 1$ for $r \le \sqrt{2}$ and $T_r < 1$ for $r > \sqrt{2}$, vibration isolation can be achieved only in the range of $r > \sqrt{2}$.
2. Since damping increases transmissibility in the isolation range ($r > \sqrt{2}$), the most effective vibration absorbers consist of spring elements having little or no damping. This implies a tradeoff when selecting a soft spring to reduce the transmitted force but increases the static displacement or vice versa with a stiff spring.

Displacement transmitted to foundations can also be determined by considering the relative displacement, $u = u^t - u_g$, which can be written as

$$\frac{u_o}{u_{go}} = \frac{r^2}{\sqrt{(1-r^2)^2 + (2r\zeta)^2}} \tag{3.57}$$

EXAMPLE 3.14

A 50-lb package is suspended in a box (damping of 2%) by two springs each having a stiffness k of 250 lb/in. The box is transported on the bed of a truck, which due to the suspension motions experiences vertical harmonic excitations of amplitude $u(t) = 1.5$ in sin (2 rad/s · t). The owner of the box has specified that the maximum relative displacement of the package be <0.05 in. Does the system satisfy this requirement?

SOLUTION

There are two ways to solve this problem; we can directly determine the relative displacement or use transmissibility to determine the total displacement and then compute the relative displacement. In both cases, we first need the frequency ratio.

1. Determine the frequency ratio of the idealized SDOF system. Mass,

$$m = W/g = 50 \text{ lbs}/(386.4 \text{ in/s}^2) = 0.1294 \text{ lbs} \cdot \text{s}^2/\text{in}$$

Stiffness,

$$k_t = 2k = 2(250 \text{ lbs/in}) = 500 \text{ lbs/in}$$

Natural frequency,

$$\omega_n = \sqrt{\frac{k_t}{m}} = \sqrt{\frac{500}{0.1294}} = 62.16 \text{ rad/s}$$

Also, the forcing frequency is given in the displacement equation as

$$\omega = 2 \text{ rad/s}$$

So the frequency ratio is

$$r = \frac{\omega}{\omega_n} = \frac{2}{62.16} = 0.0322$$

2. Determine the relative displacement of the mass. Using the transmissibility ratio definition of Equation 3.57, the relative displacement of the system can be determined. u_{go} is the amplitude of the system displacement given in the displacement equation as 1.5 in,

$$\frac{u_o}{u_{go}} = \frac{r^2}{\sqrt{(1-r^2)^2 + (2r\zeta)^2}}$$

$$\Rightarrow u_o = \frac{u_{go}r^2}{\sqrt{(1-r^2)^2 + (2r\zeta)^2}} = \frac{(1.5)(0.0322)^2}{\sqrt{(1-(0.0322)^2)^2 + \left(2(0.0322)(0.05)\right)^2}} = 0.00155'' < 0.05''$$

The displacement limit is satisfied!
The second approach utilizes the transmissibility relationship in Equation 3.55, where $u_{go} = 1.5$ in again!

$$T_r = \frac{u_o^t}{u_{go}} = \frac{\sqrt{1+(2r\zeta)^2}}{\sqrt{(1-r^2)^2 + (2r\zeta)^2}}$$

$$\Rightarrow u_o^t = \frac{u_{go}\sqrt{1+(2r\zeta)^2}}{\sqrt{(1-r^2)^2 + (2r\zeta)^2}} = \frac{(1.5)\sqrt{1+\left(2(0.0322)(0.05)\right)^2}}{\sqrt{(1-(0.0322)^2)^2 + (2(0.0322)(0.05))^2}} = 1.50155''$$

Here the relative displacement is given by

$$u_o = u_o^t - u_{go} = 1.50155'' - 1.5'' = 0.00155''$$

Same as above!

3.3 BASE SHEARS AND STRESSES CAUSED BY TIME-DEPENDENT FORCES AND SUPPORT EXCITATIONS

After determining the maximum dynamic displacement, either from a dynamic excitation or force, we can conduct a static structural analysis to determine element forces (bending moment, shear force, and axial force) and stresses needed for design; no additional dynamic analysis is necessary. For the simple portal frame system that we have considered in this chapter, we can follow one of two different approaches to conduct this structural analysis:

1. The dynamic effect can be replaced with a slowly applied force that produces deformation $u_{max} = u_o$ (or acceleration $\ddot{u}_{max} = \ddot{u}_o$), which is determined by a dynamic analysis. As shown in the prior two sections, this displacement can equally be produced by a dynamically applied force or a support excitation. Displacements can be used to determine an equivalent, external static force, $F_{s,max}$, as shown in Figure 3.23, which can in turn be used to determine all internal forces.
2. With known dynamic displacements at column ends, the forces (moment and shear) can be determined using individual element stiffness via fixed-end moment analyses, see a structural analysis textbook. Figure 3.24 depicts two cases we have used in this chapter. The first case shows the base shear and moment, which are also equal to the maximum internal shear and moment. The second case has maximum shear and moment at the top and bottom of the column.

In both cases, maximum stresses (shear and flexural) can be determined using standard strength of materials procedures. For example, the maximum shear stress is

$$\tau = \frac{VQ}{It}$$

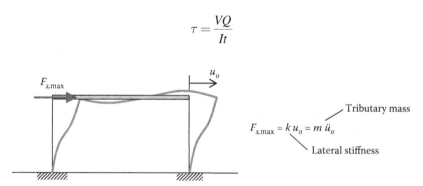

FIGURE 3.23 Equivalent static force.

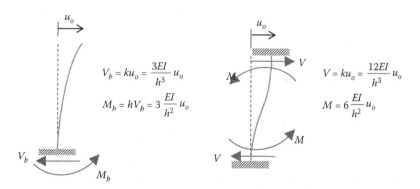

FIGURE 3.24 Internal shear force and moment for pinned–fixed and fixed–fixed columns.

where:

Q is the first moment of the area above the plane in question
I is the moment of inertia of the cross section
t is the width of the section at the plane in question

This equation yields the following maximum stresses for rectangular and I-shaped sections:
Rectangular section,

$$\tau_{max} = 1.5 \frac{V_{max}}{A} \tag{3.58}$$

I-shaped section,

$$\tau_{max} \approx \frac{V_{max}}{A_{web}} \tag{3.59}$$

where:

A is the area of the entire rectangular cross section
A_{web} is the area of the web; in practice, this is equal to the product of the depth of section and thickness of the web

The maximum flexural stress can be determined as

$$\sigma_{max} = \frac{M_{max}}{S} \tag{3.60}$$

where:

S is the section elastic modulus

EXAMPLE 3.15

Given the building frame shown in Figure E3.17 with damping of 5% and load $p(t) = 200$ lbs sin 5.3 t, determine (a) maximum displacement u, (b) the maximum base shear, and (c) the maximum normal stresses in the columns ($I_x = 75$ in⁴ and $S_x = 18$ in³ for W8 × 21). Let us assume that the beam is rigid.

SOLUTION

1. Determine mass, stiffness, and natural frequency of the SDOF system.

   The building frame can be modeled as an SDOF system assuming that the horizontal beam is rigid and only lateral deformations of the columns occur. The stiffness of the system is then the sum of column lateral stiffnesses. The mass and stiffness of the SDOF system are calculated as follows:

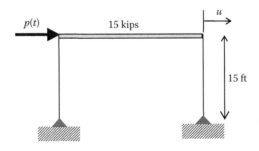

FIGURE E3.17   Structural frame and loading.

Mass,

$$m = \frac{W}{g} = \frac{15\,\text{kips}(1000\,\text{lb/kip})}{386\,\text{in/s}^2} = 38.86\,\text{lb}\cdot\text{s}^2/\text{in}$$

Stiffness,

$$k = \frac{3EI}{L^3} + \frac{3EI}{L^3} = 2\left[\frac{3(29,000,000\,\text{psi})(75\,\text{in}^4)}{(15\,\text{ft}\times12\,\text{in/ft})^3}\right] = 2237.6\,\text{lb/in}$$

The natural frequency of the SDOF system is determined using Equation 3.2:

$$\omega_n = \sqrt{\frac{k}{m}} = \sqrt{\frac{2237.6\,\text{lb/in}}{38.86\,\text{lb}\cdot\text{s}^2/\text{in}}} = 7.59\,\text{rad/s}$$

2. Determine the maximum steady-state displacement.

The maximum steady-state displacement, u_o, due to the applied forcing function can be determined using the definition of the deformation response factor in Equation 3.47.

The equivalent static displacement can be calculated using the basic force–displacement relationship for the SDOF system as follows:

$$(u_{st})_o = \frac{p_o}{k} = \frac{200\,\text{lb}}{2237.6\,\text{lb/in}} = 0.08938\,\text{in}$$

Also, given forcing frequency of $\omega = 5.3$ rad/s, the ratio of forcing frequency to natural frequency can be determined as follows:

$$r = \frac{\omega}{\omega_n} = \frac{5.3}{7.59} = 0.6984$$

Now substituting this value into Equation 3.47, the maximum steady-state displacement is

$$u_o = \frac{(u_{st})_o}{\sqrt{(1-r^2)^2 + (2r\zeta)^2}} = \frac{0.08938\,\text{in}}{\sqrt{(1-0.698^2)^2 + (2(0.698)(0.05))^2}} = 0.1729\,\text{in}$$

3. Maximum base shear.

Calculate maximum base shear at the support for each column using the force–displacement relationship:

$$V_{max} = ku_o = \left(\frac{3EI}{L^3}\right)u_o = \left[\frac{3(29,000,000\,\text{psi})(75\,\text{in}^4)}{(15\,\text{ft}\times12\,\text{in/ft})^3}\right]0.1729\,\text{in} = 193.5\,\text{lb}$$

4. Maximum normal stress due to bending.

The maximum bending moment at the top of the column is calculated from equilibrium of the column member. As the column has a pin support, summing moments about the top of the column will result in the following maximum internal bending moment at the top of the column (see Figure 3.24):

$$M_{max} = V_{max}L = 193.5\,\text{lb}(15\,\text{ft}\times12\,\text{in/ft}) = 34,822\,\text{lb}\cdot\text{in}$$

Applying the flexure formula from mechanics of materials (Equation 3.60), the normal stress in the column due to the bending moment is determined as

$$\sigma_{max} = \frac{M_{max}}{S} = \frac{34{,}822\ \text{lb} \cdot \text{in}}{17\ \text{in}^3} = 2048\ \text{psi}$$

PROBLEMS

3.1 Determine the effective weight at the roof and floor levels for a 60 ft by 150 ft, three-story office building with equal story heights of 10 ft and the following loading:
Roof DL = 20 psf
Roof LL = 20 psf
Floor DL = 20 psf
Floor LL = 60 psf
Wall DL = 10 psf
    Also, the partition load is 10 psf as the occupancy is designated as office building and the floor live load is <80 psf.

3.2 The bridge beam depicted below supports a rigid deck that weighs 200 kips. Assuming that the weight is lumped at midspan, determine the natural frequency and period of the system ($E = 29{,}000$ kip/in^2 and $I_x = 13{,}000$ in^4).

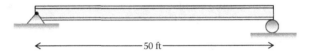

3.3 Given a water tank supported on a slender ($I = 10{,}000$ in^4), concrete ($E = 3 \times 10^7$ psi) column as shown, determine the natural frequency and period of the system.

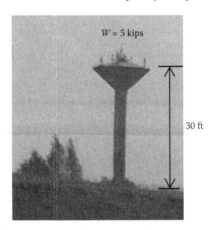

3.4 Given the following steel ($E = 29{,}000$ kip/in^2) building frame, determine the period. The columns have $I_x = 75$ in^4 and the beam has $I_x = 150$ in^4.

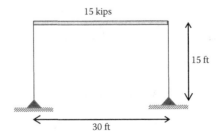

3.5 The structural system depicted below has a roof weighing 22.5 psf and side sheathing weighing 10 psf. The total weight is assumed to be concentrated at the bottom of the roof trusses. The lateral force-resisting system in the North–South direction (one on each side) consists of bracing with ½-inch steel rods ($E = 29,000$ ksi). Determine the natural period of the building.

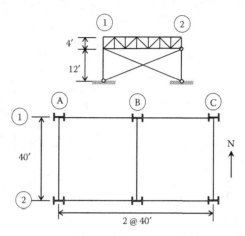

3.6 Given the following structural system and properties, determine the total stiffness and structural period.

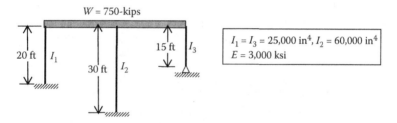

3.7 Consider the following building frame with a rigid beam that is pin connected to one column and rigidly connected to the other as shown; each column has $EI = 40,000,000$ kip-in². Determine the natural frequency and period of the system.

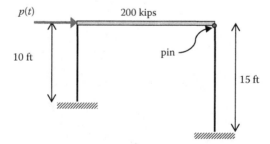

3.8 After constructing a bridge pier, the owner decides to determine the dynamic properties of the system. This type of system can be modeled as a concentrated mass atop a weightless tower. To determine the properties, a cable is attached to the top and pulled with a crane with a force of 20 kips, which causes a horizontal displacement of 1 in. The cable then is suddenly cut and the resulting free vibration is recorded. At the end of 10 complete cycles, the time is 2 s and the amplitude is 0.14 in. From this information, compute: (a) undamped natural period, (b) effective stiffness, (c) effective weight, and (d) effective damping coefficient.

3.9 A wind turbine can be modeled as a concentrated mass atop a weightless tower. To determine the dynamic properties of the system, a cable is attached to the turbine and a crane. A lateral force of 200 pounds is applied to the turbine causing a horizontal displacement of 1.0 in. When the cable is suddenly cut, the resulting free vibration is recorded. At the end of four complete cycles, the time is 1.25 s and the amplitude is 0.64 in. From this information, compute the: (a) undamped natural period, (b) effective stiffness, (c) effective weight, and (d) effective damping coefficient.

3.10 To determine the dynamic properties of a simply supported bridge girder, a cable is attached to its midspan and pulled with a crane. To produce a 4.0-inch deflection at midspan, the crane exerts a vertical force of 10 kips. When the cable is suddenly cut, the resulting free vibration is recorded. At the end of five complete cycles, the time is 0.25 s and the amplitude is 2.7 in. From this information, compute: (a) undamped natural period, (b) effective stiffness, (c) effective weight, and (d) effective damping coefficient.

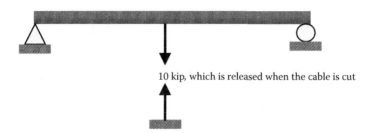

10 kip, which is released when the cable is cut

3.11 A vibration isolation block weighing 2000 lbs is to be installed in a laboratory on four springs so that the vibration from adjacent factory operations (as high as 1500 cycles per minute) will not disturb certain experiments. Determine the stiffness of system, k, such that the motion of the block is limited to 10% of the floor vibration; neglect damping.

3.12 A sensitive instrument that weighs 250 lb is mounted in the New Engineering Building in one of the labs. The instrument currently experiences a vibration amplitude of 0.4 in when an air compressor runs at a frequency of 35 Hz. Neglecting damping, determine (a) the stiffness necessary for a mount in order to achieve 80% isolation and (b) the resulting vibration amplitude of the instrument.

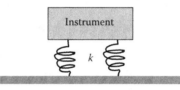

3.13 A 500-pound tiger that was rescued will be placed inside a box and transported on a flatbed truck. The box will be supported by four springs, each having stiffness of 250 lb/in. The damping for the box supporting system has been estimated at 2%. During the transport process, it is estimated that the box will experience vertical displacements of $u(t) = 1.5$ in $\cdot$ sin 4 rad/s $\cdot$ t. For the tiger to have a comfortable trip, maximum relative displacement and velocity of the box should be <0.05 in and 1.0 in/s, respectively. Does the spring damping system satisfy these requirements?

3.14 Given the following building frame with damping of 5% and load $p(t) = 200$ lbs sin(5.3 rad/s $\cdot$ t), determine (a) maximum displacement u, (b) the maximum base shear, and (c) the maximum normal stresses in the columns ($I_x = 75$ in^4 and $S_x = 18$ in^3 for W8 $\times$ 21). Assume that the beam is rigid.

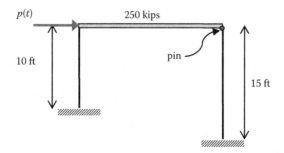

3.15 The following frame has a rigid beam, a diagonal 1/2-in diameter steel rod ($E = 29,000$ ksi) brace, and two steel columns ($E = 29,000$ ksi and $I_x = 82.7$ in^4). Determine the maximum force in the diagonal member and the maximum shear force in each of the columns when the frame is subjected to force $p(t) = 900$ lbs sin 8 rad/s · t. Damping is 5%.

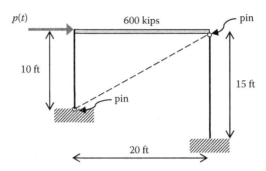

REFERENCES

American Society of Civil Engineers, ASCE/SEI *Minimum Design Loads for Buildings and Other Structures*, Edition, American Society of Civil Engineers, Reston, VA, 2010.

Chopra, A. K., *Dynamics of Structures: Theory and Applications to Earthquake Engineering*, 4th edition, Prentice-Hall, Upper Saddle River, NJ, 2012.

Lindeburg, M. R. and K. M. McMullin, *Seismic Design of Building Structures: A Professional Introduction to Earthquake Forces and Design Details*, 9th edition, Professional Publications, Inc., Belmont, CA, 2008.

Silva, P. and S. S. Badie, Optimum beam-to-column stiffness ratio of portal frames under later loads. *Structure Magazine*, August 2008.

Villaverde, R., *Fundamental Concepts of Earthquake Engineering*, CRC Press, Boca Raton, FL, 2009.

4 Response to General Loading

After reading this chapter, you will be able to:

1. Determine the response of a single degree of freedom (SDOF) system to a general load by the direct and numerical integration of Duhamel's integral
2. Use shock spectra to solve general dynamics load problems
3. Perform the direct numerical solution of equation of motion to solve for the response of an SDOF system subjected to a general dynamic load

Thus far, we covered the response of structures to harmonic loadings; however, the ground vibration of a structure during an earthquake is nonperiodic, and it can be of short or long duration. The response of an SDOF system subjected to a nonperiodic force can be determined using a convolution integral, Laplace transform, or numerical methods. The convolution integral and Laplace transform approaches are analytical methods that can be used to evaluate the response of the system in a closed form when the nonperiodic forcing function can be described analytically. The analytical approaches are useful in describing the response of an SDOF system and characterizing the influence of parameters on system response, as well as informing system design. Numerical methods are often employed in cases where an analytical solution is difficult or not available under a general forcing function.

In this chapter, we first focus on the use of a convolution integral known as Duhamel's integral in the context of structural dynamics to analyze an SDOF system subjected to a general forcing function. Duhamel's integral is based on the response of the system due to an impulse, which is a force that abruptly jumps from zero to a constant value. Several examples using Duhamel's integral are outlined in order to demonstrate the analytical solution to nonperiodic, general forcing functions; these include step, rectangular pulse, and ramp forcing functions. These analytical examples are used to introduce the concept of the shock spectra for a given forcing function and how shock spectra can be used to determine the maximum response of structures to such loadings.

In addition, we demonstrate the use of numerical methods to evaluate the response of an SDOF system to a seismic ground motion. In particular, we provide examples of how an earthquake response spectrum is generated for a given earthquake event.

4.1 RESPONSE OF AN SDOF SYSTEM TO AN IMPULSE

In this section, we determine the response of an SDOF system to a nonperiodic forcing function using Duhamel's integral. A nonperiodic forcing function, or a general forcing function, is characterized by the magnitude of the force, which varies with time and occurs during a time period of specified length. An impulsive force represents the most basic nonperiodic forcing function. An impulse is the effect of a force on a structure over a very short time span. Figure 4.1 illustrates an impulse loading function, $p(t)$, that is applied to an SDOF system and shows the impulse calculated as the area of the shaded rectangle. The impulse represents a force that acts on the mass of a structure for a specific interval of time, $\Delta\tau$.

Using Newton's second law of motion, the equation of motion can be written in terms of velocity $\dot{u}$. Notice that we only include the effects of mass because damping and stiffness have negligible effect on the maximum response for impulsive loadings since there is little time for mass to experience a displacement.

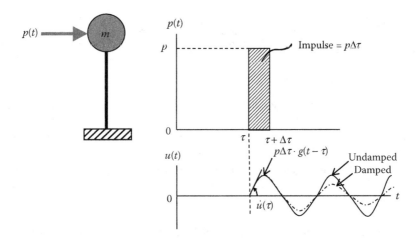

FIGURE 4.1 Impulse loading applied to an SDOF system and response.

For the equilibrium of mass, we obtain:

$$\sum F = ma : p(t) = m\frac{d\dot{u}}{dt} \tag{4.1}$$

where:

$p(t)$ represents the general loading function

Rearranging Equation 4.1 and integrating the result over the limits of $t_1 = \tau$ to $t_2 = \tau + \Delta\tau$ result in the principle of linear impulse and momentum (the product of mass and velocity), which directly relates a change in momentum $md\dot{u}$ to the impulse applied to the SDOF system $p(t)dt$, that is, the magnitude of an impulse is equal to a change in momentum,

$$md\dot{u} = p(t)dt \tag{4.2}$$

Integrating,

$$md\dot{u} = \int_{t_1}^{t_2} p(t)dt = \int_{\tau}^{\tau + \Delta\tau} p(t)dt = p\Delta\tau = \text{impulse} \tag{4.3}$$

Solving for the change in velocity yields

$$d\dot{u} = \frac{p\Delta\tau}{m} \tag{4.4}$$

Since the impulse acts for an infinitesimally short duration of time ($\Delta\tau$ is very small), the spring and damper have no effect during the time the impulse is applied. The impulse essentially imparts a velocity to the mass of the system. The change in velocity described in Equation 4.4 is equal to the initial velocity of the system, since prior to time $t = \tau$, the system was at rest.

Considering an underdamped SDOF system, subjected to the impulse at time $t = \tau$ as shown in Figure 4.1, the equation of motion is given by

$$m\ddot{u} + c\dot{u} + ku = 0 \tag{4.5}$$

where: (like in Chapter 3)
   m is the mass
   c is the damping coefficient
   k is the stiffness of the system

The initial conditions can now be defined using the initial velocity on the mass at time $t = \tau$, based on Equation 4.4. The initial velocity is given by

$$\dot{u}(\tau) = \frac{p\Delta\tau}{m} \tag{4.6}$$

and displacement before and up to the impulse at time $t = \tau$ is zero,

$$u(\tau) = 0 \tag{4.7}$$

The solution to the equation of motion is given by the free vibration response of an underdamped SDOF system shown in Equation 3.36 and rewritten in the following for an impulse applied at time $t = \tau$:

$$u(t) = e^{-\zeta\omega_n(t-\tau)}\left[\left(\frac{\dot{u}(\tau) + \zeta\omega_n u(\tau)}{\omega_D}\right)\sin\omega_D(t-\tau) + u(\tau)\cos\omega_D(t-\tau)\right] \tag{4.8}$$

Substituting the initial conditions from Equations 4.6 and 4.7 into Equation 4.8, the response of an underdamped SDOF subjected to an impulse at time $t = \tau$ is as follows:

$$u(t) = e^{-\zeta\omega_n(t-\tau)}\left[\left(\frac{p\Delta\tau}{m\omega_D}\right)\sin\omega_D(t-\tau)\right] \tag{4.9}$$

Neglecting damping in the system (because the short impulse does not allow the damping mechanism to absorb much energy), the response function is given by

$$u(t) = \left(\frac{p\Delta\tau}{m\omega_n}\right)\sin\omega_n(t-\tau) \tag{4.10}$$

The impulse response function of a damped and undamped SDOF system is shown in Figure 4.1.
If a unit impulse is applied to the system, such that $p\Delta\tau = 1$, then the response of the SDOF system becomes

$$g(t-\tau) \equiv u(t) = e^{-\zeta\omega_n(t-\tau)}\left[\left(\frac{1}{m\omega_D}\right)\sin\omega_D(t-\tau)\right] \tag{4.11}$$

Here, we define $g(t - \tau)$ as the unit impulse response function. The unit impulse response can be multiplied by the magnitude of the impulse in order to ascertain the response of the system for damped and undamped cases, as shown in Figure 4.1, $u(t) = p\Delta\tau \cdot g(t - \tau)$.

4.2   GENERAL FORCING FUNCTION

A general loading function given by $p(t)$, as shown in Figure 4.2, can be represented by a series of short impulses having varying magnitudes at successive incremental times, one after another. At some time, τ the force $p(\tau)$ acts on the system for a short duration of time $\Delta\tau$. The impulse acting on the system at time $t = \tau$ is given as $p(\tau)\Delta\tau$, and at any time t, the elapsed time since the pulse is $t - \tau$. Each impulse causes an incremental displacement response $\Delta u(t)$ of the SDOF system. The incremental displacement response to an impulse at time $t = \tau$ is given by multiplying the impulse $p(\tau)$ $\Delta\tau$ and the unit impulse response function $g(t - \tau)$, as shown in Figure 4.1, and can be rewritten as

$$\Delta u(t) = [p(\tau)\Delta\tau]g(t-\tau) \quad \text{for} \quad t > \tau \tag{4.12}$$

By summing the incremental displacement responses, the total displacement response from all the incremental impulses can be determined,

$$u(t) = \sum \Delta u(t) \cong \sum [p(\tau)\Delta\tau]g(t-\tau) \tag{4.13}$$

As $\Delta\tau$ becomes infinitesimally small or approaches zero, the summation can be replaced with integration, yielding the following result:

$$u(t) = \int_0^t p(\tau)g(t-\tau)d\tau \tag{4.14}$$

Then substituting the definition of the impulse response function from Equation 4.11 into Equation 4.14, we can determine the total response of an underdamped SDOF system at any time t as the superposition of the response to all the impulses occurring before t, that is,

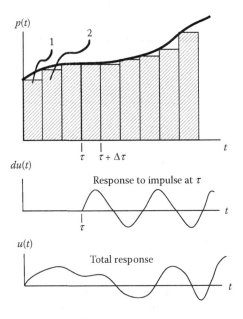

FIGURE 4.2   Discretization of a general loading function and response.

$$u(t) = \frac{1}{m\omega_D} \int_0^t p(\tau) e^{-\zeta\omega_n(t-\tau)} [\sin \omega_D(t-\tau)] d\tau \qquad (4.15)$$

Equation 4.15 is a convolution integral and is known as Duhamel's integral. It allows for the determination of the response of an underdamped SDOF system to any general forcing function, and is particularly useful for determining the effect of random excitations associated with ground motion during earthquakes. Also, if the forcing function $p(t)$ cannot be expressed by an analytical equation, which is often the case in earthquake-induced ground motion, the integral of Equation 4.15 can be evaluated using numerical methods. It is important to recognize that Equation 4.15 assumes that the system is at rest before the application of the forcing function $p(t)$. Also, Equation 4.15 accounts for the particular solution of the response $u_p(t)$ as described in Chapter 3, with the total solution given by Equation 3.23. If an initial displacement $u(0)$ and initial velocity $\dot{u}(0)$ were present, and the complementary (free vibration response) part of the solution given by Equation 3.36 would be added to Equation 4.15, then the total response of an underdamped case including initial conditions is

$$u(t) = e^{-\zeta\omega_n t} \left[\left(\frac{\dot{u}(0) + \zeta\omega_n u(0)}{\omega_D} \right) \sin \omega_D t + u(0)\cos \omega_D t \right]$$
$$+ \frac{1}{m\omega_D} \int_0^t p(\tau) e^{-\zeta\omega_n(t-\tau)} \left[\sin \omega_D(t-\tau) \right] d\tau \qquad (4.16)$$

If damping in the system is neglected, then Equations 4.15 and 4.16 can be expressed as follows:

$$u(t) = \frac{1}{m\omega_n} \int_0^t p(\tau)[\sin \omega_n(t-\tau)] d\tau \qquad (4.17)$$

and

$$u(t) = \left(\frac{\dot{u}(0)}{\omega_n} \right) \sin \omega_n t + u(0)\cos \omega_n t + \frac{1}{m\omega_n} \int_0^t p(\tau)[\sin \omega_n(t-\tau)] d\tau \qquad (4.18)$$

Again, it is important to note that Equation 4.17 assumes that the system is initially at rest, whereas Equation 4.18 is given to show how the initial displacement, $u(0)$, and the initial velocity, $\dot{u}(0)$, would be included in the response.

EXAMPLE 4.1

Consider the undamped SDOF system, shown in Figure 4.1, subjected to the constant force shown in Figure E4.1. Assuming that the SDOF system is initially at rest, use Duhamel's integral to determine the displacement response, $u(t)$, of the system.

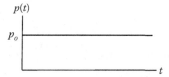

FIGURE E4.1   Step or constant force loading function.

SOLUTION

1. Describe the forcing function and equation of motion to be solved. The forcing function $p(t)$ can be described analytically as

$$p(t) = p_o$$

This forcing function can be considered a step force, or a dynamically applied constant force, because it is abruptly applied to the mass of the system as opposed to a force that is applied very slowly inducing no vibration on the system and can be considered statically applied.

In this problem, damping is neglected and so the equation of motion to be solved is given as

$$m\ddot{u}(t) + ku(t) = p_o$$

2. Apply Duhamel's integral. Since we are neglecting damping effects in the system ($\zeta = 0$) and the system is initially at rest ($u(0) = 0$ and $\dot{u}(0) = 0$), Equation 4.17 can be applied directly. Substituting the forcing function into the equation and integrating results in the response of the system,

$$u(t) = \frac{1}{m\omega_n} \int_0^t p_o \sin\omega_n(t - \tau)d\tau = \frac{p_o}{m\omega_n^2} \cos\omega_n(t - \tau)\Big|_0^t$$

or

$$u(t) = \frac{p_o}{m\omega_n^2}[1 - \cos\omega_n t]$$

Recalling that stiffness $k = \omega_n^2 m$, the response can be written as follows:

$$u(t) = \frac{p_o}{k}[1 - \cos\omega_n t] = u_{st}[1 - \cos\omega_n t]$$

where:

u_{st} represents the static displacement of the SDOF system; that is, if the force p_o were applied very slowly such that no vibrations were induced, then the displacement would be $u_{st} = p_o/k$

The term $[1 - \cos\omega_n t]$ is the dynamic load factor (DLF) and represents the amplification of the displacement above the static displacement; that is, $u(t)/u_{st}$.

3. Plot the response of the SDOF system. By rearranging the response in terms of the DLF and substituting for $\omega_n = 2\pi/T_n$, the following equation can be used to represent the response of the SDOF system:

$$\frac{u(t)}{u_{st}} = DLF = \left[1 - \cos 2\pi\frac{t}{T_n}\right]$$

We can use this equation to obtain the displacement response of the SDOF system subjected to a step force (Figure E4.2). The response of an SDOF system to the step force $p(t) = p_o$ is similar to the free vibration response of an undamped system except that it is shifted up by the static displacement u_{st}. The maximum displacement ($u_{max} = 2u_{st}$) of an SDOF system subjected to a suddenly applied force p_o is twice the displacement of

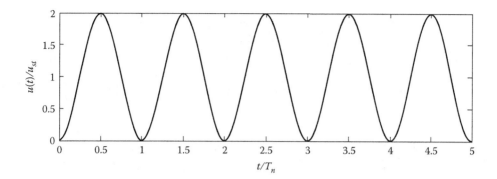

FIGURE E4.2 Response of an undamped SDOF to step force.

the same SDOF system when an equivalent static force p_o is applied very slowly, and it occurs at $\omega t = \pi$ or when $t = T/2$.

EXAMPLE 4.2

Use Duhamel's integral to determine the response of an undamped SDOF system subjected to a rectangular impulse that is applied from $0 < t < t_d$. The rectangular impulse is a suddenly applied constant force, p_o, which is applied for a specified duration of time, t_d, as shown in Figure E4.3. Assume that the SDOF system is initially at rest.

SOLUTION

1. Describe the forcing function and equation of motion to be solved. The forcing function $p(t)$ can be described analytically as

$$p(t) = \begin{cases} p_o & t \le t_d \\ 0 & t > t_d \end{cases}$$

By superposition, the given forcing function can be described as a sum of step functions where the first forcing function $p_1(t)$, with a magnitude of p_o, starts at $t = 0$, and the second forcing function $p_2(t)$, with a magnitude of $-p_o$, starts at $t = t_d$. This is analytically described as

$$p_1(t) = p_o$$

$$p_2(t) = \begin{cases} 0 & t \le t_d \\ -p_o & t > t_d \end{cases}$$

and graphically illustrated using superposition as the sum of step forces shown in Figure E4.4.

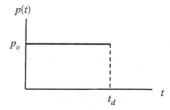

FIGURE E4.3 Rectangular impulse loading function.

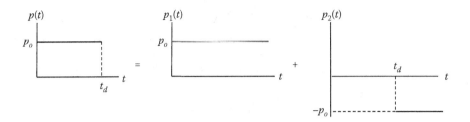

FIGURE E4.4 Illustration of the rectangular impulse as the sum of two step force loading functions.

2. Apply Duhamel's integral. The response of the SDOF system to the step force $p_1(t) = p_o$ can be determined by applying Duhamel's integral and substituting the forcing function. This result is the same as the response determined in Example 4.1 and rewritten in the following for convenience:

$$u_1(t) = u_{st}[1 - \cos \omega_n t]$$

where:

$u_1(t)$ is the response of the SDOF system to the first forcing function $p_1(t)$

The second forcing function $p_2(t)$ is a time-delayed step force with a magnitude of $-p_o$, and the response to this second forcing function can also be determined by substituting the forcing function and evaluating Duhamel's integral as follows:

$$u_2(t) = \frac{1}{m\omega_n} \int_{t_d}^{t} -p_o \sin \omega_n(t - \tau) d\tau = \left. \frac{-p_o}{m\omega_n^2} \cos \omega_n(t - \tau) \right|_{t_d}^{t}$$

$$u_2(t) = \frac{-p_o}{k}[1 - \cos \omega_n(t - t_d)] = u_{st}[\cos \omega_n(t - t_d) - 1]$$

There are two response regions due to the rectangular pulse: during the pulse when $t \le t_d$, the total response of the system is given by $u_1(t)$, and after the pulse when $t > t_d$, the total response of the system is given by the two responses, $u_1(t) + u_2(t)$, that is,

$$u(t) = \begin{cases} u_1(t) & t \le t_d \\ u_1(t) + u_2(t) & t > t_d \end{cases} \quad \text{or} \quad u(t) = \begin{cases} u_{st}[1 - \cos \omega_n t] & t \le t_d \\ u_{st}[\cos \omega_n(t - t_d) - \cos \omega_n t] & t > t_d \end{cases}$$

which can be written in terms of the DLF = $u(t)/u_{st}$ and a dimensionless time parameter, the ratio of duration of the loading to the natural period of the structure, t_d/T_n:

$$\text{DLF} = \frac{u(t)}{u_{st}} = \begin{cases} \left[1 - \cos 2\pi \dfrac{t}{T_n} \right] & t \le t_d \\ \left[\cos 2\pi \left(\dfrac{t}{T_n} - \dfrac{t_d}{T_n} \right) - \cos 2\pi \dfrac{t}{T_n} \right] & t > t_d \end{cases}$$

3. Plot the response of the SDOF system. We can use these equations for different values of t_d/T_n to obtain the displacement response of the SDOF system subjected to a rectangular pulse force as shown in Figure E4.5.

Although we can find closed-form solutions to some general forcing function problems, it is more efficient to determine the results using computer software, such as MATLAB®. In this chapter, we use MATLAB to develop short scripts to solve many of the problems encountered in structural

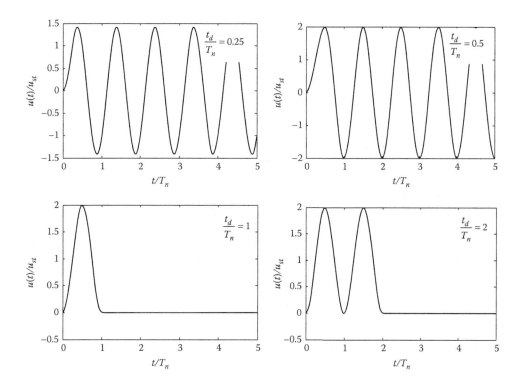

FIGURE E4.5 Response of an SDOF system due to a rectangular pulse and varying pulse durations.

dynamics. And in later chapters, we will use these MATLAB scripts to solve earthquake engineering problems. These programs have powerful algorithms that can perform complex calculations efficiently. In fact, modern analysis and design have become highly dependent on these computational tools.

MATLAB (MATrix LABoratory) was originally designed as an interactive software system for matrix operations, such as solving systems of linear equations, and computing eigenvalues and eigenvectors. More recent versions of the program include extensive graphics capabilities that can be used to easily plot results, such as those shown in Examples 4.1 and 4.2. It also has an extensive library of functions, including the one that performs convolution integrals numerically (the *conv* function), as well as differentiation and integration. To facilitate the development of a script to perform the convolution operation, we first rewrite Equation 4.17 (or Equation 4.15) in terms of two functions that are then discretized into dimensionless vectors. The first function includes the shape of the pulse and is written as a generalized forcing function $p(t) = p_o x(t)$ that can be substituted into Equation 4.17 (as $p(\tau) = p_o x(\tau)$),

$$u(t) = \frac{p_o}{m\omega_n} \int_0^t x(\tau) \cdot \sin \omega_n (t - \tau) d\tau \qquad (4.19)$$

We also write this equation in terms of a dimensionless time factor, the ratio of time to the natural period of the structure, t/T_n. That is, substituting $\omega_n = 2\pi/T_n$ and $\omega_n = \sqrt{k/m}$, we obtain

$$u(t) = \frac{\omega_n}{\omega_n} \frac{p_o}{m\omega_n} \int_0^t x(\tau) \cdot \sin \omega_n (t - \tau) d\tau = \frac{2\pi}{T_n} \frac{p_o}{m\left(\sqrt{k/m}\right)^2} \int_0^t x(\tau) \cdot \sin \frac{2\pi}{T_n} (t - \tau) d\tau$$

or

$$\frac{u(t)}{u_{st}} = 2\pi \int_0^t x\left(\frac{\tau}{T_n}\right) \cdot \sin\left[2\pi\left(\frac{t}{T_n} - \frac{\tau}{T_n}\right)\right]d\tau \tag{4.20}$$

where $u_{st} = p_o/k$. Equation 4.20 can be written in convolution form as

$$\text{DLF} = \frac{u(t)}{u_{st}} = 2\pi \int_0^t x\left(\frac{\tau}{T_n}\right) \cdot h\left[2\pi\left(\frac{t}{T_n} - \frac{\tau}{T_n}\right)\right]d\tau \tag{4.21}$$

where the function $h(\xi) = \sin(\xi)$, and ξ is the argument of the function, which is the quantity in the brackets in this case.

EXAMPLE 4.3

Use the convolution function in MATLAB and Equation 4.21 to solve for the response due to the rectangular impulse shown in Figure E4.3.

SOLUTION

$$x(t) = \begin{cases} 1 & t \le t_d \\ 0 & t > t_d \end{cases}$$

1. The MATLAB script is as follows:

```
clear all
clc
% specify the duration of the pulse td/T
tdT = input ('Enter the value for td/T ratio: ');
% create a vector with n equally spaced points to represent t/Tn
n = 500;
tT = linspace(0,5,n);% loop over t/Tn to create vector of input pulse
for i = 1:length(tT)
    if tT(i) <= tdT
        p(i) = 1;
    else
        p(i) = 0;
    end
end
% integrate pulse function to get response
dt=tT(2)-tT(1);
p=p*dt;
h=sin(2*pi*tT);
uust=2*pi*conv(p,h);
%create plot
plot (tT, uust(1:length(tT)), 'LineWidth',2, 'Color',[0 0 0]);
xlabel ('t/T_n', 'FontSize', 12, 'FontName', 'Times New Roman',…
    'FontAngle', 'italic');
ylabel ('u(t)/u_{st}', 'FontSize', 12, 'FontName', 'Times New Roman', …
    'FontAngle', 'italic');
```

This script can be used to generate graphs of the normalized displacement response of SDOF systems subjected to a rectangular impulse, identical to those obtained using the results of Example 2.

EXAMPLE 4.4

Use the convolution function of MATLAB and Equation 4.21 to determine graphically the dynamic response of a tower subjected to the load shown in Figure E4.6.

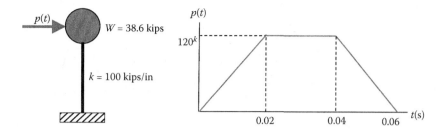

FIGURE E4.6 An SDOF system and trapezoidal impulse loading function.

SOLUTION

1. Mass, stiffness, and natural frequency of the SDOF system. The tower can be modeled as an SDOF system. The stiffness of the system is given as 100 kips/in and the mass can be calculated as follows:

$$m = \frac{W}{g} = \frac{38.6 \text{ kips}(1000 \text{ lb/kip})}{386.4 \text{ in/s}^2} = 100 \text{ lb s}^2/\text{in}$$

The natural frequency of the SDOF system is determined using Equation 3.2,

$$\omega_n = \sqrt{\frac{k}{m}} = \sqrt{\frac{100,000}{100}} = 31.6 \text{ rad/s}$$

2. Obtain function $x(t)$:

$$x(t) = \begin{cases} 50t & t \le 0.02 \text{ s} \\ 1 & 0.02 \text{ s} < t \le 0.04 \text{ s} \\ 3 - 50t & 0.04 \text{ s} < t \le 0.06 \text{ s} \\ 0 & t > 0.06 \text{ s} \end{cases}$$

Also, $p_o = 120$ kips
3. Displacement response as a function of time script:

```
clear all
clc
% input data
omega = 31.6; % natural frequency in rad/sec
po=120000; % peak force in lbs
m=100; % mass in lbs per square second
dt=0.002; % time increment for calculations
% loop over t to create vector based on pulse function values
for i = 1:500;
    t(i)=dt*i;
    if t(i) <= 0.02
        p(i) = 50*t(i);
    elseif t(i)<= 0.04
        p(i) = 1;
    elseif t(i)<= 0.06
        p(i) = 3-50*t(i);
    else
        p(i) = 0;
    end
end
```

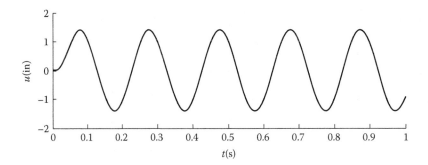

FIGURE E4.7   Response of an SDOF system to trapezoidal loading function.

```
p=p*dt;
h=sin(omega*t);
uust=po*conv(p,h)/(m*omega);
uustr = max(abs(uust)) % maximum displacement
%create plot
plot (t, uust(1:500), 'LineWidth',2, 'Color',[0 0 0]);
xlabel ('t(sec)', 'FontSize', 12, 'FontName', 'Times New Roman',...
    'FontAngle', 'italic');
ylabel ('u(in)', 'FontSize', 12, 'FontName', 'Times New Roman', ...
    'FontAngle', 'italic');
```

The maximum dynamic displacement is 1.3968 in versus the static displacement of $p_o/k = 1.2$ in (Figure E4.7).

4.3   SHOCK SPECTRA

The results of Example 4.2 (Figure E4.5) show that each time we specify a different natural period of the structure relative to the duration of the pulse (t_d/T_n), we obtain a new response, each of which has a maximum value. In design, only these peak values are necessary and not the entire history of the response. The envelope of the maximum response as a function of the loading duration relative to the period of the structure can be graphically represented as a *shock spectrum*. The shape of the envelope, or the shock spectrum, is also dependent on the shape of the pulse. The response used to obtain shock spectra can be expressed in terms of displacement, velocity, or acceleration. This is useful in determining the values of the peak (absolute) response that are used in design.

To obtain shock spectra, we can solve Duhamel's integral analytically or numerically, both of which were introduced in the previous section. Numerical integration can also be performed using one of the standard numerical integration techniques such as Euler (forward rectangular), Trapezoidal (linear), or Simpson's (parabolic) methods, all of which can be found in a standard calculus textbook. In this book, we use the numerical integration operators available in MATLAB; for Duhamel's integral, we apply the *conv* function used in Example 4.3. Following are some examples used to generate the shock spectra using both analytical solutions and convolution numerical integration. Also, we demonstrate the use of these spectra in determining the maximum response of structural systems.

EXAMPLE 4.5

Use the results from Example 4.2 to draw the shock spectrum for the rectangular pulse force, as shown in Figure E4.2.

SOLUTION

1. Use the DLF versus t_d/T_n results obtained in part 2 of Solution 4.2; that is,

$$\text{DLF} = \frac{u(t)}{u_{st}} = \begin{cases} \left[1 - \cos 2\pi \dfrac{t}{T_n}\right] & t \le t_d \\[3mm] \left[\cos 2\pi \left(\dfrac{t}{T_n} - \dfrac{t_d}{T_n}\right) - \cos 2\pi \dfrac{t}{T_n}\right] & t > t_d \end{cases}$$

The maximum values of the DLF, DLF_{max}, for various values of t_d/T_n can be obtained more efficiently using MATLAB. The results of DLF_{max} versus t_d/T_n can then be graphed to obtain the shock spectrum of a pulse loading, as shown in the Figure E4.8. The script is as follows:

```
clear all
clc
% create two vectors of equal n length to represent td/Tn and t/Tn
n = 500;
tdT = linspace(0,5,n);
tT = linspace(0,5,n);
% loop over td/Tn and t/Tn to create the shock spectrum
for j = 1:n
for i = 1:n
    if tT(i) <= tdT(j)
        uust(i) = 1 - cos(2*pi*tT(i));
    else
        uust(i) = cos(2*pi*(tT(i)-tdT(j)))- cos(2*pi*tT(i));
    end
end
uustr(j)=max(abs(uust)); % Select the max values from each response
end
% create figure
figure1=figure;
axes1 = axes('Parent',figure1);
xlim (axes1,[0 5]);
ylim (axes1,[-2.5 2.5]);
box (axes1, 'on');
```

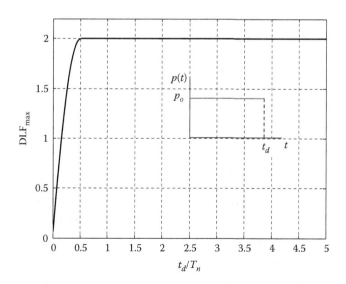

FIGURE E4.8 Shock spectrum of an SDOF system subject to a rectangular impulse load.

```
        grid (axes1, 'on');
        hold (axes1, 'all');
        %create plot of the shock spectrum
        plot (tdT,uustr,'LineWidth',2,'Color',[0 0 0]);
        xlabel ('t_d /T_n','FontSize',12,'FontName','Times New Roman',...
            'FontAngle','italic');
        ylabel ('DLF_{max}','FontSize',12,'FontName','Times New Roman',...
            'FontAngle','italic');
```

EXAMPLE 4.6

Write a MATLAB script using the *conv* function to plot the shock spectrum for the rectangular pulse in Example 4.2.

SOLUTION

```
clear all
clc
% create two vectors of equal n length to represent td/Tn and t/Tn
n = 500;
tdT = linspace(0,5,n);
tT = linspace(0,5,n);
% loop over td/Tn and t/Tn to create the shock spectrum
for j = 1:n
    for i = 1:n
        if tT(i) <= tdT(j)
            p(i) = 1;
        else
            p(i) = 0;
        end
    end
    % integrate pulse function to get response
    dt=tT(2)-tT(1);
    p=p*dt;
    h=sin(2*pi*tT);
    uust=2*pi*conv(p,h);
    uustr(j)=max(abs(uust)); % Select the max values from each response
end
% create figure
figure1=figure;
axes1 = axes('Parent',figure1);
xlim (axes1,[0 5]);
ylim (axes1,[0 2.2]);
box (axes1, 'on');
grid (axes1, 'on');
hold (axes1, 'all');
%create plot of the shock spectrum
plot (tdT,uustr,'LineWidth',2,'Color',[0 0 0]);
xlabel ('t_d /T_n','FontSize',12,'FontName','Times New Roman',...
    'FontAngle','italic');
ylabel ('DLF_{max}','FontSize',12,'FontName','Times New Roman',...
    'FontAngle','italic');
```

This script creates a graph identical to the graph shown in Figure E4.8. The difference in the two scripts is in the way they establish the two arrays that are plotted to generate the figure. Example 4.5 uses the analytical solution, while Example 4.6 applies the *conv* function to obtain the response array.

EXAMPLE 4.7

Draw the shock spectrum for the response of an SDOF system subjected to the symmetric triangular pulse loading function shown in Figure E4.9.

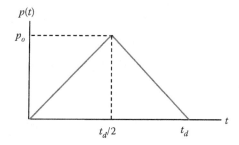

FIGURE E4.9 Symmetric triangular impulse loading function.

<small>**SOLUTION**</small>

1. Obtain function $x(t)$:

$$x(t) = \begin{cases} 2t/t_d & t \le t_d/2 \\ 2 - 2t/t_d & t_d/2 < t \le t_d \\ 0 & t > t_d \end{cases}$$

2. Displacement response as a function of time script: (Figure E4.10)

```
clear all
clc
% create two vectors of equal n length to represent td/Tn and t/Tn
n = 500;
tdT = linspace(0,5,n);
tT = linspace(0,5,n);
% loop over td/Tn and t/Tn to create the shock spectrum
for j = 1:n
    for i = 1:n
        if tT(i) <= tdT(j)/2
            p(i) = 2*tT(i)/tdT(j);
        elseif tT(i) <= tdT(j)
            p(i) = 2-2*tT(i)/tdT(j);
```

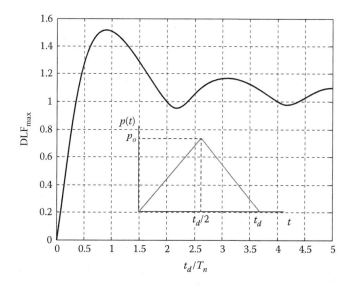

FIGURE E4.10 Shock spectrum of an SDOF system subject to a symmetric triangular impulse load.

```
        else
            p(i) = 0;
        end
    end
    % integrate pulse function to get response
    dt=tT(2)-tT(1);
    p=p*dt;
    h=sin(2*pi*tT);
    uust=2*pi*conv(p,h);
    uustr(j)=max(abs(uust)); % Select the max values from each response
end
% create figure
figure1=figure;
axes1 = axes('Parent',figure1);
xlim (axes1,[0 5]);
ylim (axes1);
box (axes1, 'on');
grid (axes1, 'on');
hold (axes1, 'all');
%create plot of the shock spectrum
plot (tdT,uustr,'LineWidth',2,'Color',[0 0 0]);
xlabel ('t_d /T_n','FontSize',12,'FontName','Times New Roman',...
    'FontAngle','italic');
ylabel ('DLF_{max}','FontSize',12,'FontName','Times New Roman',...
    'FontAngle','italic');
```

EXAMPLE 4.8

Draw the shock spectrum for the response of an SDOF system subjected to the triangular pulse loading function shown in Figure E4.11.

SOLUTION

1. Obtain function $x(t)$:

$$x(t) = \begin{cases} 1-t/t_d & t \le t_d \\ 0 & t > t_d \end{cases}$$

2. Displacement response as a function of time script (Figure E4.12):

```
clear all
clc
% create two vectors of equal n length to represent td/Tn and t/Tn
n = 500;
```

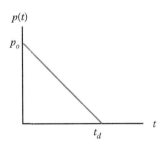

FIGURE E4.11   Triangular impulse loading function.

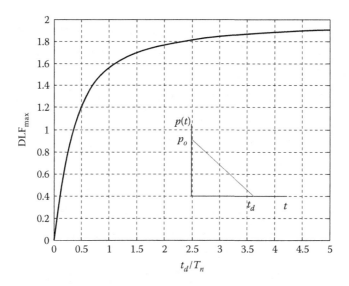

FIGURE E4.12 Shock spectrum of an SDOF system subject to a triangular impulse load.

```
tdT = linspace(0,5,n);
tT = linspace(0,5,n);
% loop over td/Tn and t/Tn to create the shock spectrum
for j = 1:n
    for i = 1:n
        if tT(i) <= tdT(j)
            p(i) = 1-tT(i)/tdT(j);
        else
            p(i) = 0;
        end
    end
    % integrate pulse function to get response
    dt=tT(2)-tT(1);
    p=p*dt;
    h=sin(2*pi*tT);
    uust=2*pi*conv(p,h);
    uustr(j)=max(abs(uust)); % Select the max values from each response
end
% create figure
figure1=figure;
axes1 = axes('Parent',figure1);
xlim (axes1,[0 5]);
ylim (axes1);
box (axes1, 'on');
grid (axes1, 'on');
hold (axes1, 'all');
%create plot of the shock spectrum
plot (tdT,uustr,'LineWidth',2,'Color',[0 0 0]);
xlabel ('t_d /T_n','FontSize',12,'FontName','Times New Roman',…
    'FontAngle','italic');
ylabel ('DLF_{max}','FontSize',12,'FontName','Times New Roman',…
    'FontAngle','italic');
```

EXAMPLE 4.9

Draw the shock spectrum for the response of an SDOF system subjected to the half-cycle sine pulse force shown in Figure E4.13.

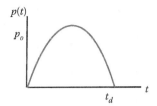

FIGURE E4.13   Half-cycle sine impulse loading function.

SOLUTION

1. Obtain function $x(t)$:

$$x(t) = \begin{cases} \sin(\pi t/t_d) & t \leq t_d \\ 0 & t > t_d \end{cases}$$

2. Displacement response as a function of time script (Figure E4.14):

```
clear all
clc
% create two vectors of equal n length to represent td/Tn and t/Tn
n = 500;
tdT = linspace(0,5,n);
tT = linspace(0,5,n);
% loop over td/Tn and t/Tn to create the shock spectrum
for j = 1:n
    for i = 1:n
        if tT(i) <= tdT(j)
            p(i) = sin(pi*tT(i)/tdT(j));
        else
            p(i) = 0;
        end
    end
end
% integrate pulse function to get response
dt=tT(2)-tT(1);
```

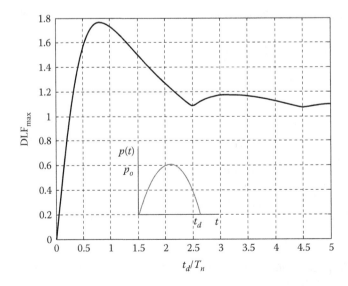

FIGURE E4.14   Shock spectrum of an SDOF system subject to a half-cycle sine impulse load.

```
        p=p*dt;
        h=sin(2*pi*tT);
        uust=2*pi*conv(p,h);
        uustr(j)=max(abs(uust)); % Select the max values from each response
    end
    % create figure
    figure1=figure;
    axes1 = axes('Parent',figure1);
    xlim (axes1,[0 5]);
    ylim (axes1);
    box (axes1, 'on');
    grid (axes1, 'on');
    hold (axes1, 'all');
    %create plot of the shock spectrum
    plot (tdT,uustr,'LineWidth',2,'Color',[0 0 0]);
    xlabel ('t_d /T_n','FontSize',12,'FontName','Times New Roman',...
        'FontAngle','italic');
    ylabel ('DLF_{max}','FontSize',12,'FontName','Times New Roman',...
        'FontAngle','italic')
```

EXAMPLE 4.10

Given the building frame shown in Figure E4.15, which is subjected to a triangular impulse force (see Figure E4.11) of amplitude $p_o = 5$ kips and duration $t_d = 0.6$ s, determine (1) maximum displacement at the top, (2) the maximum base shear, and (3) the maximum bending stresses in the columns ($I_x = 82.7$ in⁴ and $S_x = 20.9$ in³ for W8 × 24). Assume that the beam is rigid.

SOLUTION

1. Mass, stiffness, and natural period and frequency of the SDOF system. The building frame can be modeled as an SDOF system, assuming that only lateral deformations of the columns occur. The stiffness of the system is the sum of column lateral stiffnesses. The mass and stiffness of the SDOF system are calculated as follows:

    Mass:

$$m = \frac{W}{g} = \frac{20 \text{ kips}(1000 \text{ lb/kip})}{386.4 \text{ in/s}^2} = 51.8 \text{ lb s}^2/\text{in}$$

Stiffness (see Table 3.1); the 15 ft column is fixed–fixed, while the 20 ft column is fixed–pinned,

$$k = \frac{12EI}{h_1^3} + \frac{3EI}{h_2^3} = \frac{12(29,000,000 \text{ psi})(82.7 \text{ in}^4)}{(15 \text{ ft} \times 12 \text{ in/ft})^3} + \frac{3(29,000,000 \text{ psi})(82.7 \text{ in}^4)}{(20 \text{ ft} \times 12 \text{ in/ft})^3} = 5455 \text{ lb/in}$$

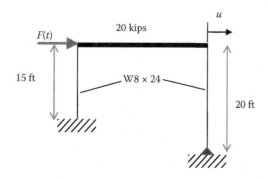

FIGURE E4.15   SDOF building frame geometry.

The natural frequency of the SDOF system is determined using Equation 3.2,

$$\omega_n = \sqrt{\frac{k}{m}} = \sqrt{\frac{5455}{51.8}} = 10.3 \text{ rad/s}$$

The natural period of the SDOF system is determined using Equation 3.5,

$$T_n = \frac{2\pi}{\omega_n} = \frac{2\pi}{10.3 \text{ rad/s}} = 0.61 \text{ s}$$

2. Determine the maximum displacement. The maximum displacement due to the applied triangular pulse forcing function can be determined using the shock spectrum obtained in Figure E4.12. This figure provides the maximum dynamic load factor, $\text{DLF}_{max} = u_o/u_{st}$ as a function of the ratio of pulse duration to natural period, t_d/T_n, where the equivalent static displacement u_{st} can be calculated using the basic force–displacement relationship for the SDOF system as follows:

$$u_{st} = \frac{p_o}{k} = \frac{5 \text{ kips}(1000 \text{ lb/kip})}{5455 \text{ lb/in}} = 0.92 \text{ in}$$

Also, the ratio of pulse duration to natural period can be determined as follows,

$$\frac{t_d}{T_n} = \frac{0.6}{0.61 \text{ s}} = 0.98$$

Finally, from the shock spectrum obtained in Figure E4.12, we can determine $\text{DFL}_{max} \cong 1.55$. Now, solve for u_{max},

$$u_o = (\text{FDL})_{max} \cdot u_{st} = 1.55(0.92 \text{ in}) = 1.42 \text{ in}$$

3. Maximum base shear. Calculate the maximum base shear at the support for each column using the force–displacement relationship or the equilibrium of the column member as described in Figure 3.24:
   The 15-foot column:

$$V_{1max} = k_1 u_o = \left(\frac{12EI}{h_1^3}\right) u_o = \left(\frac{12(29,000 \text{ ksi})(82.7 \text{ in}^4)}{(15 \text{ ft} \times 12 \text{ in/ft})^3}\right) 1.42 \text{ in} = 7.01 \text{ kips}$$

   The 20-foot column:

$$V_{2max} = k_2 u_o = \left(\frac{3EI}{h_2^3}\right) u_o = \left(\frac{3(29,000 \text{ ksi})(82.7 \text{ in}^4)}{(20 \text{ ft} \times 12 \text{ in/ft})^3}\right) 1.42 \text{ in} = 0.74 \text{ kips}$$

4. Maximum normal stress due to bending. The maximum bending moment at the top of each column is calculated from the equilibrium of the column member as described in Figure 3.24.
   The 15-foot column:

$$M_{1max} = \left(\frac{6EI}{h_1^2}\right) u_o = \left(\frac{6(29,000 \text{ ksi})(82.7 \text{ in}^4)}{(15 \text{ ft} \times 12 \text{ in/ft})^2}\right) 1.42 \text{ in} = 631 \text{ kip} \cdot \text{in}$$

The 20-foot column:

$$M_{2max} = \left(\frac{3EI}{h_2^2}\right)u_o = \left(\frac{3(29,000 \text{ ksi})(82.7 \text{ in}^4)}{(20 \text{ ft} \times 12 \text{ in/ft})^2}\right)1.42 \text{ in} = 177 \text{ kip} \cdot \text{in}$$

The maximum moment yields the maximum bending stress; thus, the 15-foot column experiences the maximum stress. Applying the flexure formula from mechanics of materials (Equation 3.60), the normal stress in this column due to the bending moment is

$$\sigma_{max} = \frac{M_{1max}}{S} = \frac{631 \text{ kip} \cdot \text{in}}{20.9 \text{ in}^3} = 30.2 \text{ ksi}$$

4.4 RESPONSE TO GROUND MOTION

The equation of motion for ground excitation was derived in Chapter 3 (Equation 3.50) and is shown here for convenience:

$$m\ddot{u}(t) + c\dot{u}(t) + ku(t) = -m\ddot{u}_g(t) \tag{4.22}$$

The total response of an underdamped SDOF system including initial conditions subjected to a ground excitation, $\ddot{u}_g(t)$, can be determined using Equation 4.16 with a new forcing function

$$p(t) = -m\ddot{u}_g(t) \tag{4.23}$$

When the system starts from rest, we can determine the response due to a ground excitation for an underdamped system in terms of the relative displacement of the mass with respect to the ground by substituting Equation 4.22 into Duhamel's integral (Equation 4.15). That is,

$$u(t) = -\frac{1}{\omega_D}\int_0^t \ddot{u}_g(t)e^{-\zeta\omega_n(t-\tau)}[\sin\omega_D(t-\tau)]\,d\tau \tag{4.24}$$

EXAMPLE 4.11

Draw the response of the frame shown in Figure E4.15, when subjected to the ground motion of the North–South component of the horizontal ground acceleration recorded at the Imperial Valley Irrigation District substation, El Centro, California, during the Imperial Valley Earthquake of May 18, 1940, from here on referred to as the El Centro earthquake. Assume 2% damping. The input ground acceleration was obtained from http://www.vibrationdata.com/elcentro.dat, and is graphically shown in Figure E4.16.

SOLUTION

1. The natural frequency of the frame ω_n was obtained in Example 4.10 as 10.3 rad/s. Using the *conv* function introduced in Section 4.2, we can write the following script to perform the analysis:

```
clear all
clc
dam = 0.02; % given damping of 2%
wn = 10.3; % rad/sec, determined in Example 10
load elcentro.mat; % load El Centro data
N = length(elcentro); % number of points in the ground acceleration file
```

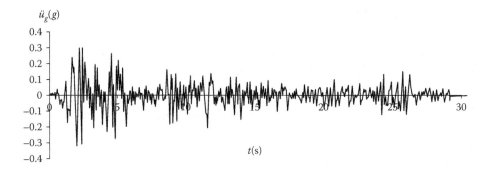

FIGURE E4.16 Ground motion acceleration time history for El Centro earthquake.

```
acc = elcentro; % ground acceleration data
DT = 0.02; % sampling rate
% create the time vector using the sampling rate
for i=N:-1:1
  t(i)=i*DT;
end
% Integrate the ground acceleration using convolution
p=acc*DT;
h=exp(-dam*wn*t).*sin(wn*t);
u=conv(p,h)*386.4/wn;
uo=max(abs(u)) % determine maximum displacement
% create a new time vector adjusted to the conv length to graph response
for j=1:2*N-1
  t2(j)=j*DT/2;
end
% create figure
figure1=figure;
axes1 = axes('Parent',figure1);
xlim (axes1,[0 20]);
ylim (axes1);
hold (axes1, 'all');
%create plot
plot (t2, u, 'LineWidth',2, 'Color',[0 0 0]);
xlabel ('t(sec)','FontName','Times New Roman','FontAngle','italic');
ylabel ('u(in)','FontName','Times New Roman','FontAngle', 'italic');
```

This gives a maximum displacement of 3.12 in, and the graphical response history results are shown in Figure E4.17.

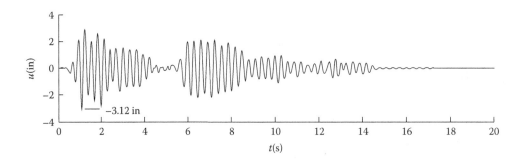

FIGURE E4.17 Displacement response of the SDOF building frame due to El Centro earthquake.

Equation 4.23 can be rewritten as the DLF in terms of the relative displacement u_o/u_{st} or the total accelerations $\ddot{u}_o^t/\ddot{u}_{go}$. Additionally, we can normalize time with respect to the natural period as a dimensionless time parameter, t/T_n. First, let $u_{st} = p_{eff}/k$; assuming $p_{eff} = m\ddot{u}_{go}$ leads to $u_{st} = m\ddot{u}_{go}/k = \ddot{u}_{go}/\omega_n^2$, since $k = \omega_n^2 m$ and $\ddot{u}_{go}$ is the peak value of the ground acceleration time history, $\ddot{u}_g$. The DLF can then be written as

$$\text{DLF} = \frac{u_o}{u_{st}} = \frac{\omega_n^2 u_o}{\ddot{u}_{go}} = \frac{\ddot{u}_o^t}{\ddot{u}_{go}} \tag{4.25}$$

where $\ddot{u}_o^t$ is the total acceleration, which is related to the relative displacement by $\omega_n^2 u_o$. From Equation 4.21, we determine the convolution integral component of Equation 4.24,

$$\text{DLF} = 2\pi \int_0^t x\left(\frac{\tau}{T_n}\right) \cdot h\left[2\pi\left(\frac{t}{T_n} - \frac{\tau}{T_n}\right)\right] d\tau \tag{4.26}$$

where $x(\xi)$ is a function that describes the applied displacement (or acceleration) history function. This implies that all of the shock spectra derived thus far are applicable to ground excitation forcing functions as well.

However, generating a shock spectrum using the *conv* function in MATLAB is relatively simple. For instance, in the following example, we use the El Centro earthquake ground acceleration presented in Example 4.11. This data will be used extensively in the remainder of the book. Also, the terminology changes when dealing with seismic response; rather than shock spectrum, it is customary (and more appropriate) to refer to the shock spectrum as the *response spectrum*. The response spectrum is the plot of the maximum response as a function of the SDOF system period. This topic will be covered extensively in Chapter 5, including the development of the design response spectrum for elastic and inelastic structural systems.

EXAMPLE 4.12

Draw the response spectrum for the response of an SDOF system subjected to the El Centro earthquake ground acceleration.

SOLUTION

The input ground acceleration can be obtained from http://www.vibrationdata.com/elcentro.dat as a fraction of the acceleration due to gravity, g. In this case, we derive the response based on Equation 4.23 and the natural period since we do not have a lapse time for the input forcing function. Thus, the displacement response as a function of the natural period script is:

```
clear all
clc
dam = 0.02; % damping; can changed to generate different spectra
load elcentro.mat; % load El Centro data
N = length(elcentro); % number of points in the ground acceleration file
acc = elcentro; % ground acceleration data
DT = 0.02; % sampling rate
for i=N:-1:1
t(i)=i*DT;
end
for j = 500:-1:1
    period(j)=j*DT/2; % natural period array
    wn(j)=2*pi/period(j); % natural frequency array
    p=acc*DT;
    h=exp(-dam*wn(j)*t).*sin(wn(j)*t);
    u=conv(p,h);
    uustr(j)=max(abs(u))/wn(j); % maximum displacement
```

```
end
%create plot
plot (pcriod, uustr*386.4, 'LineWidth',2, 'Color',[0 0 0]);
xlabel ('T_n(sec)','FontName','Times New Roman','FontAngle','italic');
ylabel ('u(in)','FontName','Times New Roman','FontAngle','italic');
```

As mentioned previously, seismic ground accelerations cannot be described analytically and the evaluation of the integral in Equation 4.23 typically requires the use of numerical methods in order to characterize the response of the SDOF system subjected to ground accelerations. Numerical integration using the *conv* function in MATLAB, or other numerical integration methods, can be difficult to implement for complex cases, and are not applicable in the case of nonlinear behavior. Therefore, seismic ground excitation loadings are typically treated using the direct integration of the equation of motion. Following is a discussion of three direct integration methods.

4.5 DIRECT INTEGRATION METHODS

Since the numerical integration of the convolution (Duhamel's) integral is difficult to implement for complex loading functions, and not applicable for nonlinear response, time-stepping numerical methods for the direct integration of the differential equation of motion were developed. First, to derive any direct integration numerical algorithm, we assume an SDOF system characterized by the natural frequency ω_n and the damping ratio ζ. Also, we divide the acceleration excitation $a(t)$ function into N equal intervals of Δt.

Time-stepping methods are divided into two categories: *explicit* and *implicit*. Explicit methods use the state of the system at the current time to calculate the state of the system at the next time interval, while implicit methods use relationships involving both current and later states of the system to find a solution. Consequently, implicit methods require iteration, making them more difficult to implement. However, implicit methods are more practical in cases where the use of an explicit method requires impractically small time steps, Δt, to keep the error in the result bounded. Thus, for a desired level of accuracy, implicit methods can take much less computational time because they can use longer time steps. Summaries of three of the most commonly applied methods follow.

4.5.1 Nigam–Jennings Algorithm (Explicit)

For each interval Δt, the response is calculated using the conditions at the beginning of the time interval—initial conditions. These initial conditions are given by the displacement and velocity at the end of the preceding time interval. We also assume the forcing function to be linear and piecewise continuous (see Figure 4.3),

$$a(\tau) = a_i + \left(\frac{\Delta a_i}{\Delta t}\right)\tau, \quad t_i \le \tau \le t_{i+1} \tag{4.27}$$

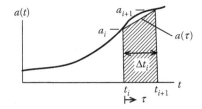

FIGURE 4.3 Discretization of acceleration time history function.

where:

$\Delta a_i = a_{i+1} - a_i$

$t_i = i \cdot \Delta t$

and

$\Delta t = t_{i+1} - t_i$

for $i = 1, 2, 3, \ldots, N$

The equation of motion for the time interval can be written as

$$\ddot{u} + 2\omega_n \zeta \dot{v} + \omega_n^2 u = -a_i - \left(\frac{\Delta a_i}{\Delta t}\right)\tau, \quad t_i \leq \tau \leq t_{i+1} \tag{4.28}$$

The solution to this linear differential equation with constant coefficients is given by the complementary and particular components,

$$u(t) = u_c(t) + u_p(t) \tag{4.29}$$

where, over the interval Δt, the complementary solution is

$$u_c(\tau) = e^{-\zeta \omega_n (\tau - t_i)}(C_i \cos \omega_D(\tau - t_i) + D_i \sin \omega_D(\tau - t_i))$$

and over the same interval, the particular solution is

$$u_p(\tau) = B_i + A_i(\tau - t_i)$$

Here, A_i and B_i are the constants of integration, determined by substituting u_p back into the equation of motion. After applying boundary conditions, the formulas to calculate the displacement, velocity, and acceleration at time step $t_{i+1} = t_i + \Delta t$ are

$$u_{i+1} = a_{11}u_i + a_{12}\dot{u}_i + b_{11}a_i + b_{12}a_{i+1} \tag{4.30}$$

$$\dot{u}_{i+1} = a_{21}u_i + a_{22}\dot{u}_i + b_{21}a_i + b_{22}a_{i+1} \tag{4.31}$$

$$\ddot{u}_{i+1} = -\omega_n^2 u_{i+1} - 2\zeta\omega_n\dot{u}_{i+1} \tag{4.32}$$

respectively. Where the coefficients of the displacements and velocities in the first two equations only need to be computed once and are given as

$$a_{11} = e^{-\zeta\omega_n\Delta t}\left(\frac{\zeta}{\sqrt{1-\zeta^2}}\sin\omega_D\Delta t + \cos\omega_D\Delta t\right)$$

$$a_{12} = e^{-\zeta\omega_n\Delta t}\frac{\sin\omega_D\Delta t}{\omega_D}$$

$$a_{21} = -\frac{\omega_n}{\sqrt{1-\zeta^2}} e^{-\zeta\omega_n \Delta t} \sin \omega_D \Delta t$$

$$a_{22} = e^{-\zeta\omega_n \Delta t} \left(\cos \omega_D \Delta t - \frac{\zeta}{\sqrt{1-\zeta^2}} \sin \omega_D \Delta t \right)$$

$$b_{11} = e^{-\zeta\omega_n \Delta t} \left[\left(\frac{2\zeta^2 - 1}{\omega_n^2 \Delta t} + \frac{\zeta}{\omega_n} \right) \frac{\sin \omega_D \Delta t}{\omega_D} - \left(\frac{2\zeta}{\omega_n^3 \Delta t} + \frac{1}{\omega_n^2} \right) \cos \omega_D \Delta t \right] - \frac{2\zeta}{\omega_n^3 \Delta t}$$

$$b_{12} = -e^{-\zeta\omega_n \Delta t} \left[\left(\frac{2\zeta^2 - 1}{\omega_n^2 \Delta t} \right) \frac{\sin \omega_D \Delta t}{\omega_D} + \frac{2\zeta}{\omega_n^3 \Delta t} \cos \omega_D \Delta t \right] - \frac{1}{\omega_n^2} + \frac{2\zeta}{\omega_n^3 \Delta t}$$

$$b_{21} = e^{-\zeta\omega_n \Delta t} \left[\left(\frac{2\zeta^2 - 1}{\omega_n^2 \Delta t} + \frac{\zeta}{\omega_n} \right) \left(\cos \omega_D \Delta t - \frac{\zeta}{\sqrt{1-\zeta^2}} \sin \omega_D \Delta t \right) \right.$$

$$\left. - \left(\frac{2\zeta}{\omega_n^3 \Delta t} + \frac{1}{\omega_n^2} \right) \left(\omega_D \sin \omega_D \Delta t + \zeta\omega_n \cos \omega_D \Delta t \right) \right] + \frac{1}{\omega_n^2 \Delta t}$$

$$b_{22} = \frac{e^{-\zeta\omega_n \Delta t}}{\omega_n^2 \Delta t} \left(\frac{\zeta}{\sqrt{1-\zeta^2}} \sin \omega_D \Delta t + \cos \omega_D \Delta t \right) - \frac{1}{\omega_n^2 \Delta t}$$

EXAMPLE 4.13

Write a MATLAB script for the Nigam–Jennings algorithm; then use the script to draw displacement, velocity, and acceleration response spectra for the response of the SDOF system subjected to the El Centro earthquake ground acceleration presented in Example 4.11.

SOLUTION

Again, the input ground acceleration can be obtained from http://www.vibrationdata.com/elcentro.dat as a fraction of the acceleration due to gravity g. In this example, we write a MATLAB script based on Equations 4.30 and 4.31. The maximum displacement, velocity, and acceleration response as a function of the natural period script is:

```
clear all
clc
d(1) = 0;% initial displacement
v(1) = 0;% initial velocity
dam = 0.02; % damping; can be changed to generate different response spectra
DT = 0.02; % sampling rate
load elcentro.mat; % load El Centro data
N = length(elcentro); % number of points in the ground acceleration file
acc = elcentro; % ground acceleration data
% generate the data to graph a response spectrum by changing Omega
for j = 250:-1:1
period(j) = j*DT; % natural period
wd = 2*pi/period(j)*sqrt(1-dam^2); % damped frequency
wn = 2*pi/period(j); % natural frequency
```

```
%%% compute values of the response and choose the maximum %%%
% first calculate the elements of the matrices A and B
a11=exp(-dam*wn*DT)*(dam/sqrt(1-dam^2)*sin(wd*DT)+cos(wd*DT));
a12=exp(-dam*wn*DT)/wd*sin(wd*DT);
a21=-wn/sqrt(1-dam^2)*exp(-dam*wn*DT)*sin(wd*DT);
a22=exp(-dam*wn*DT)*(cos(wd*DT)-dam/sqrt(1-dam^2)*sin(wd*DT));
b11=exp(-dam*wn*DT)*(((2*dam^2-1)/(wn^2*DT)+dam/wn)*sin(wd*DT)/wd+...
(2*dam/(wn^3*DT)+1/wn^2)*cos(wd*DT))-2*dam/(wn^3*DT);
b12=-exp(-dam*wn*DT)*((2*dam^2-1)/(wn^2*DT)*sin(wd*DT)/wd+...
2*dam/(wn^3*DT)*cos(wd*DT))-1/wn^2+2*dam/(wn^3*DT);
b21=exp(-dam*wn*DT)*(((2*dam^2-1)/(wn^2*DT)+dam/wn)*(cos(wd*DT)-...
dam/sqrt(1-dam^2)*sin(wd*DT))-(2*dam/(wn^3*DT)+1/wn^2)*(wd*sin(wd*DT)+...
dam*wn*cos(wd*DT)))+1/(wn^2*DT);
b22=(-1+exp(-dam*wn*DT)*(dam/sqrt(1-dam^2)*sin(wd*DT)+cos(wd*DT)))/(wn^2*DT);
%loop to find response
for i = 1:N-1
    d(i+1) = a11*d(i)+a12*v(i)+b11*acc(i)+b12*acc(i+1);
    v(i+1) = a21*d(i)+a22*v(i)+b21*acc(i)+b22*acc(i+1);
    Responseaccelation(i) = -2*wn*dam*v(i+1)-(wn^2)*d(i+1);
end
% compute the value of the largest response
max_acc(j) = max(abs(Responseaccelation));
max_vel(j) = max(abs(v));
max_dis(j) = max(abs(d));
end
subplot(3,1,1), plot(period,max_dis*386.4)
ylabel('Displacement(in)','FontName','Times New Roman','FontAngle','italic')
subplot(3,1,2), plot(period,max_vel*386.4)
ylabel('Velocity(in/s)','FontName','Times New Roman','FontAngle','italic')
subplot(3,1,3), plot(period,max_acc)
xlabel('T_n(sec)','FontName','Times New Roman','FontAngle', 'italic')
ylabel('Acceleration(g)','FontName','Times New Roman','FontAngle','italic')
```

The results for the displacement response spectrum obtained using this method and the *conv* function in MATLAB, shown in Figure E4.18, are virtually identical. The other spectrum are shown in Figure E4.19.

4.5.2 CENTRAL DIFFERENCE METHOD (EXPLICIT)

With this method, the response is calculated using the constant time interval Δt, based on a finite difference approximation of the time derivatives of the displacement in the equation of motion. We also assume the excitation forcing function to be linear and piecewise continuous, with approximate time derivatives for time step i of duration Δt shown in Figure 4.4. Substituting the velocity

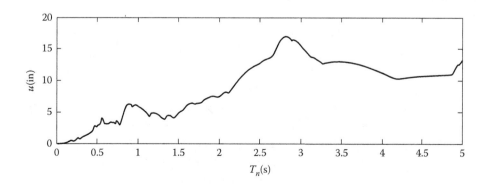

FIGURE E4.18 Displacement response spectrum for an SDOF system due to El Centro earthquake.

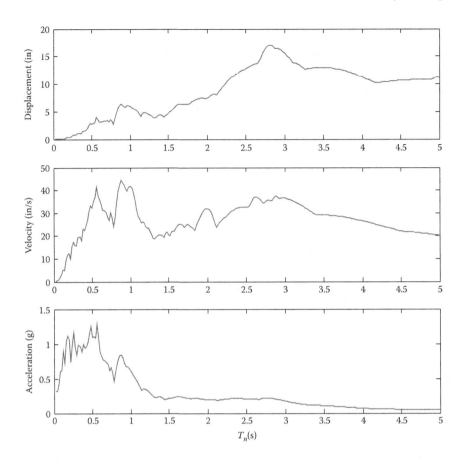

FIGURE E4.19 Displacement, velocity, and acceleration response spectra for the SDOF system subjected to El Centro earthquake.

and acceleration as shown in this figure into the equation of motion, Equation 3.49, we obtain the equation of motion for the time interval $2\Delta t$,

$$\frac{u_{i+1} - 2u_i + u_{i-1}}{(\Delta t)^2} + 2\omega_n \zeta \frac{u_{i+1} - u_{i-1}}{2\Delta t} + \omega_n^2 u_i = -\ddot{u}_{gi} \tag{4.33}$$

where $\ddot{u}_{gi}$ is the ground acceleration at time step i. Also, u_i and u_{i-1} are known from the preceding step. Rearranging Equation 4.33 to have these known quantities on the right,

$$\left(\frac{1}{(\Delta t)^2} + \frac{\omega_n \zeta}{\Delta t}\right) u_{i+1} = -\ddot{u}_{gi} - \left(\frac{1}{(\Delta t)^2} - \frac{\omega_n \zeta}{\Delta t}\right) u_{i-1} - \left(\omega_n^2 - \frac{2}{(\Delta t)^2}\right) u_i \tag{4.34}$$

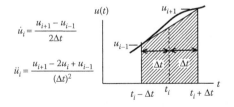

FIGURE 4.4 Finite difference approximation of velocity and acceleration.

or

$$\hat{k}u_{i+1} = \hat{a}_i \tag{4.35}$$

where:

$$\hat{k} = \frac{1}{(\Delta t)^2} + \frac{\omega_n \zeta}{\Delta t}$$

$$\hat{a}_i = -\ddot{u}_{gi} - \left(\frac{1}{(\Delta t)^2} - \frac{\omega_n \zeta}{\Delta t}\right)u_{i-1} - \left(\omega_n^2 - \frac{2}{(\Delta t)^2}\right)u_i = -\ddot{u}_{gi} - au_{i-1} - bu_i$$

Determining the initial velocity $\dot{u}_o$ and displacement u_o from the specified initial conditions, we can use the following algorithm for the central difference method:

1. Initial calculations based on the known quantities:

$$\ddot{u}_o = -\ddot{u}_{go} - 2\omega_n \zeta \dot{u}_o - \omega_n^2 u_o$$

$$u_{-1} = u_o - \Delta t \cdot \dot{u}_o + \frac{(\Delta t)^2}{2} \cdot \ddot{u}_o$$

$$\hat{k} = \frac{1}{(\Delta t)^2} + \frac{2\omega_n \zeta}{2\Delta t}$$

$$a = \frac{1}{(\Delta t)^2} - \frac{\omega_n \zeta}{\Delta t}$$

$$b = \omega_n^2 - \frac{2}{(\Delta t)^2}$$

2. Calculations for step i, where $i = 0, 1, 2, 3, \ldots$

$$\hat{a}_i = -\ddot{u}_{gi} - au_{i-1} - bu_i$$

$$u_{i+1} = \frac{\hat{a}_i}{\hat{k}}$$

$$\dot{u}_i = \frac{u_{i+1} - u_{i-1}}{2\Delta t}$$

$$\ddot{u}_i = \frac{u_{i+1} - 2u_i + u_{i-1}}{(\Delta t)^2}$$

This method is conditionally stable. It requires the following time step length for stability.

$$\Delta t < \frac{T_n}{\pi} \tag{4.36}$$

This is rarely a limitation in seismic response analyses since Δt is typically specified as 0.02 s or smaller to accurately define the ground acceleration $\ddot{u}_g$.

4.5.3 NEWMARK'S BETA METHOD FOR LINEAR SYSTEMS (IMPLICIT)

Professor Nathan Newmark developed a general method that approximates the variation of the acceleration response within each time interval. In this method, the response can be calculated using nonconstant time intervals Δt_i following approximations of the time derivatives of the displacement in the equation of motion as

$$\dot{u}_{i+1} = \dot{u}_i + [(1-\gamma)\Delta t]\ddot{u}_i + [\gamma\Delta t]\ddot{u}_{i+1} \tag{4.37}$$

$$u_{i+1} = u_i + \Delta t \cdot \dot{u}_i + [(0.5-\beta)(\Delta t)^2]\ddot{u}_i + [\beta(\Delta t)^2]\ddot{u}_{i+1} \tag{4.38}$$

Where parameters β and γ define the variation of the acceleration over nonconstant time intervals Δt_i. The values of these parameters must be chosen carefully in order to obtain stable results. Two sets of values that produce stable results include $\beta = 1/2$ and $\gamma = 1/4$ (constant average acceleration) and $\beta = \frac{1}{2}$ and $\gamma = 1/6$ (linear acceleration).

After obtaining the initial velocity $\dot{u}_o$ and displacement u_o from the specified initial conditions, we can use the following algorithm for the Newmark Beta method:

1. Initial calculations based on the known quantities and time intervals Δt_i:

$$\ddot{u}_o = -\ddot{u}_{go} - 2\omega_n\zeta\dot{u}_o - \omega_n^2 u_o$$

$$a_1 = \frac{1}{\beta(\Delta t)^2} + \frac{2\omega_n\zeta\gamma}{\beta\cdot\Delta t}$$

$$a_2 = \frac{1}{\beta\cdot\Delta t} + 2\omega_n\zeta\left(\frac{\gamma}{\beta}-1\right)$$

$$a_3 = \frac{1}{2\beta} - 1 + 2\omega_n\zeta\left(\frac{\gamma}{2\beta}-1\right)\Delta t$$

$$\hat{k} = \omega_n^2 + a_1$$

2. Calculations for step i, where $i = 0, 1, 2, 3, \ldots,$

$$\hat{a}_{i+1} = -\ddot{u}_{gi+1} + a_1 u_i + a_2 \dot{u}_i + a_3 \ddot{u}_i$$

$$u_{i+1} = \frac{\hat{a}_{i+1}}{\hat{k}}$$

$$\dot{u}_{i+1} = \frac{\gamma}{\beta\cdot\Delta t}(u_{i+1}-u_i) + \left(1-\frac{\gamma}{\beta}\right)\dot{u}_i + \Delta t\left(1-\frac{\gamma}{2\beta}\right)\ddot{u}_i$$

$$\ddot{u}_{i+1} = \frac{(u_{i+1} - u_i)}{\beta(\Delta t)^2} - \frac{\dot{u}_i}{\beta \cdot \Delta t} - \left(\frac{1}{2\beta} - 1\right)\ddot{u}_i$$

This method is also conditionally stable. It requires the following time step for stability:

$$\Delta t \le T_n \Big/ \pi \sqrt{2(\gamma - 2\beta)} \tag{4.39}$$

which for $\beta = 1/2$ and $\gamma = 1/4$ is unconditionally stable.

This particular method is useful in the analysis of nonlinear systems; however, details of the algorithm are slightly different than those presented here. For complete details of the analysis of nonlinear systems, see the textbook by Chopra (2012).

PROBLEMS

4.1 Use Duhamel's integral to determine the response of an undamped SDOF system subjected to a half-sine impulse that is applied from $0 < t < t_d$. Assume that the SDOF system is initially at rest and $t_d/T \ne 0.5$. Hint, this can be solved by using superposition of two sinusoidal excitations.

4.2 Use Duhamel's integral to determine the dynamic response of a tower subjected to the load shown.

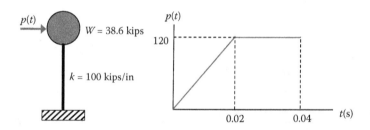

4.3 Draw the shock spectrum for an SDOF system for the triangular impulse loading shown.

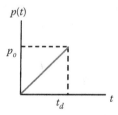

4.4 Assume that Burn's tower (a water tank weighing 4200 kips) on the campus of University of the Pacific is supported on a 120 ft tall cantilever tower with stiffness of 2000 kips/in., determine the design values for lateral deformation and base shear for a symmetric triangular impulse of amplitude 100 kips and duration $t_d = 0.1$ s.

4.5 The bridge beam depicted below supports a rigid deck that weighs 200 kips. The beam is subjected to the blast load described by the triangular impulse loading shown. Determine the maximum stress in the beam. ($E = 29{,}000$ kip/in^2, $I_x = 13{,}000$ in^4, and $S_x = 830$ in^3).

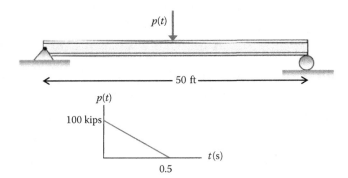

4.6 Consider the following building frame with a rigid beam that is pinned connected to one column and rigidly connected to the other as shown; each column has
$EI = 40,000,000$ kip in^2. The frame is subjected to the blast load shown at the beam
level. Determine the maximum stress in each column given $S_x = 2000$ in^3.

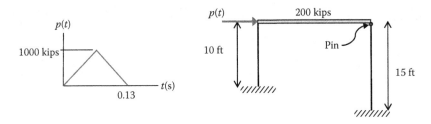

4.7 Write a MATLAB script for the central difference algorithm; then use the script to
draw displacement, velocity, and acceleration response spectra for the response of the
SDOF system subjected to the El Centro earthquake ground acceleration presented in
Example 4.11.

4.8 Write a MATLAB script for the Newmark Beta algorithm; then use the script to
draw displacement, velocity, and acceleration response spectra for the response of the
SDOF system subjected to the El Centro earthquake ground acceleration presented in
Example 4.11.

4.9 Write a MATLAB script to solve Problem 4.8 using the *lsim* function in MATLAB,
which simulates time response of dynamic systems to arbitrary inputs.

REFERENCE

Chopra, A. K., *Dynamics of Structures: Theory and Applications to Earthquake Engineering*, 4th edition,
Prentice-Hall, Upper Saddle River, NJ, 2012.

5 Response Spectrum Analysis of SDOF System

After reading this chapter, you will be able to:

1. Use response spectra to calculate the maximum response of a structure, specifically displacements and base shear
2. Obtain a design elastic response spectrum
3. Use design response spectra to calculate maximum response of a structure, specifically displacements and base shear
4. Develop an inelastic response spectrum
5. Use an inelastic response spectrum to calculate the maximum response of a structure, specifically displacements and base shear

Given the complexity of seismic ground acceleration records, numerical methods such as those introduced in Chapter 4 must be applied to obtain the response of a structure to an input ground excitation (see Figure 5.1). The response depicted in Figure 5.1 is analogous to the analysis performed in Example 4.11 of Chapter 4. For seismic-resistant design, the entire response history of the system is not necessary, only the absolute maximum value of the response is required. However, it is necessary to process a large portion of the response history to identify this maximum response value. Without the aid of computer software, this process can be time consuming, even for the simple impulsive loadings presented in Chapter 4. Also, the response (displacement, velocity, or acceleration) of a structure to a seismic excitation depends on system dynamic properties (mass, stiffness, and damping) and ground excitation characteristics. As discussed in Chapter 4, it is more convenient to use a plot of the maximum response to a specified loading for all SDOF systems based on their inherent natural period (or frequency) and damping; the so-called shock spectrum for impulse loading and response spectrum for seismic loading. Here, we review the theoretical concepts as well as a practical approach for developing a response spectrum that can be used for design; this is the precursor to the contemporary design response spectrum used in current design codes (see Chapter 8). Since most systems are allowed to deform inelastically during a seismic event, we also present details of the design response spectrum for systems in the inelastic range.

5.1 ELASTIC RESPONSE SPECTRUM

Response spectrum analysis is an effective approach to assess the behavior of an SDOF system when it is subjected to a particular peak ground acceleration (PGA) input. And as shown in Example 4.12 of Chapter 4, developing an elastic response spectrum is a straightforward process, provided we can use a computer program to perform the tedious calculations. The elastic response spectrum only depends on the system natural period and damping because these are the only dynamic properties that influence the maximum values of the response. The response may be expressed as displacement (relative or total), velocity (relative or total), or acceleration (relative or total), as shown by Example 4.13 in Chapter 4; however, the convention is to show relative displacement, relative velocity, and total acceleration.

Figures 5.2 and 5.3 illustrate the general process for developing an elastic response spectrum. First, we select an earthquake accelerogram (ground acceleration vs. time); also known as a seismic time-history. The time-history for various earthquakes (at different sites) can be obtained from a

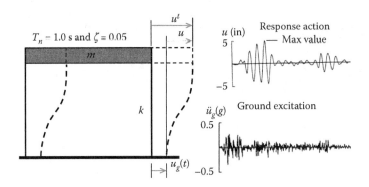

FIGURE 5.1 Response of a building frame to a seismic ground excitation.

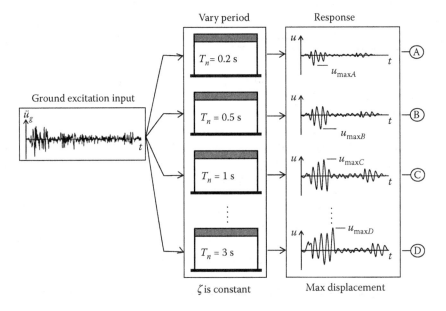

FIGURE 5.2 Displacement response to a seismic ground excitation.

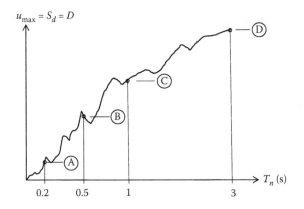

FIGURE 5.3 Displacement response spectrum from Figure 5.2 results.

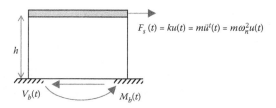

FIGURE 5.4   Equivalent static force, base shear, and overturning moment.

number of websites, particularly http://ngawest2.berkeley.edu/. For a specific seismic time-history, natural period, T_n, and damping ratio, ζ, we generate a complete response history (relative displacement, relative velocity, or total acceleration) using one of the numerical methods discussed in Chapter 4. Figure 5.2 illustrates the process for generating relative displacement responses. We then scan each response history to identify the maximum value, which corresponds to a single point on the response spectrum plot, as shown in Figure 5.3. For instance, the maximum displacement response for case A in Figure 5.2 corresponds to point A on the response spectrum in Figure 5.3. We then repeat the process for different natural periods (with constant damping) until the response spectrum plot is completed. Finally, we repeat the entire process for different values of damping to generate other response spectra.

The response quantities graphed on these spectra are known as spectral quantities; that is,

$S_d = D =$ spectral displacement = maximum relative displacement of an SDOF system.
$S_v =$ spectral velocity = maximum relative velocity of an SDOF system.
$S_a =$ spectral acceleration = maximum total acceleration of an SDOF system.

As discussed in Section 3.3, we only need the relative displacement to conduct a structural analysis (see Figure 5.4). That is, the base shear, V_b, and the overturning moment, M_b, can be determined using the static equilibrium analysis of the structure when subjected to an equivalent static force, $F_s = ku(t)$; $V_b = m\omega_n^2 u(t)$; and $M_b = hV_b$. For design purposes, we are primarily interested in the maximum response so we apply the peak values of the relative displacement or total acceleration. But the total acceleration is related to the displacement as shown in Figure 5.4. So it would appear that only the spectral displacement would be necessary to compute the peak values of deformations and internal forces needed for seismic design. However, the velocity and the acceleration response spectra are needed in order to construct design spectra and to demonstrate how these design spectra relate to modern building code specifications.

EXAMPLE 5.1

Use the MATLAB® script provided in Chapter 4, Example 4.12 to draw response spectrum of the displacement, velocity, and acceleration for an SDOF system subjected to the El Centro earthquake ground acceleration for 2%, 5%, and 10% damping.

SOLUTION

Input ground acceleration is given in Figure E5.1.

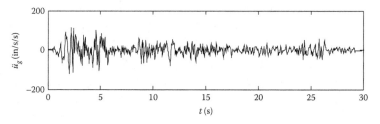

FIGURE E5.1   Seismic (acceleration) time-history for El Centro earthquake.

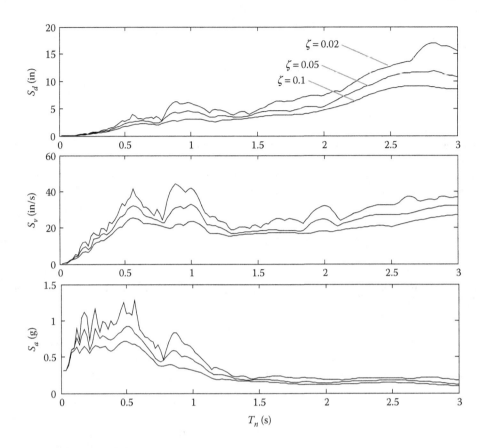

FIGURE E5.2 Displacement, velocity, and acceleration response spectra for El Centro earthquake.

The response spectra results are shown in Figure E5.2; for damping factors of 0.02, 0.05, and 0.1, top to bottom.

EXAMPLE 5.2

Use the MATLAB script provided in Chapter 4, Example 4.13 to draw the response spectrum of the displacement, velocity, and acceleration for an SDOF system subjected to the 1985 Mexico City earthquake ground acceleration for 2%, 5%, and 10% damping.

SOLUTION

Input ground acceleration is given in Figure E5.3.

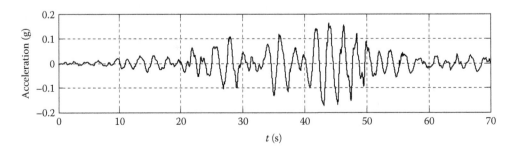

FIGURE E5.3 Seismic (acceleration) time-history for 1985 Mexico City earthquake, SCT1 (EW).

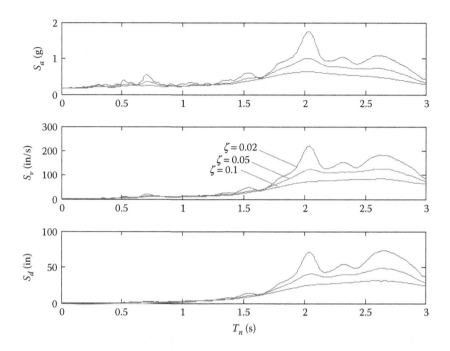

FIGURE E5.4   Seismic ground motion and response spectra for Mexico City earthquake.

The response spectra results are shown in Figure E5.4; for damping factors of 0.02, 0.05, and 0.1, top to bottom.

As can be seen from these spectra in Figure E5.4, the response is largest for a natural period of ~2 s, which is the dominant period in Mexico City as discussed in Chapter 2.

Note that the relationship between the relative displacement and total acceleration $m\ddot{u}^t(t) = m\omega_n^2 u(t)$ is only valid for systems without damping. For this reason, the acceleration determined from this relationship is known as the *pseudo-acceleration, A,* in order to distinguish A from the actual acceleration. That is, consider the equation of motion in terms of the relative quantities,

$$\ddot{u} + 2\zeta\omega_n\dot{u} + \omega_n^2 u = -\ddot{u}_g \tag{5.1}$$

Since most structures have small damping (see Table 3.3), we can assume that the velocity term is negligible ($2\zeta\omega_n\dot{u} \ll \omega_n^2 u$); thus, Equation 5.1 can be rewritten as

$$\ddot{u} + \ddot{u}_g = -\omega_n^2 u \tag{5.2}$$

Substituting the value of total acceleration, $\ddot{u} + \ddot{u}_g = \ddot{u}^t$, we obtain

$$\ddot{u}^t = -\omega_n^2 u \tag{5.3}$$

This results in the definition of the spectral *pseudo-acceleration, A,* in terms of the spectral displacement, D, as

$$A = \omega_n^2 D \tag{5.4}$$

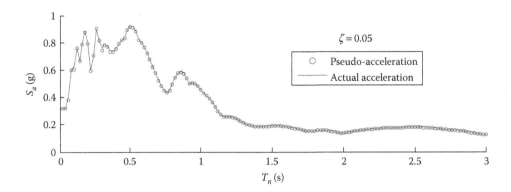

FIGURE 5.5 Comparison of actual and pseudo-acceleration response spectra for El Centro earthquake.

Note that in general $A \neq \ddot{u}^t_{\max}$, but it is adequate for most practical purposes, as shown in Figure 5.5, for 5% damping where pseudo-acceleration and actual acceleration response spectra are plotted for the El Centro earthquake. A 5% damping ratio was selected in order to match the damping level assumed in building code design spectra.

Similarly, we can derive a quantity for the relative velocity in terms of the displacement, the *pseudo-velocity, V.* For this case, we use an energy–balance relationship. Recall that the strain energy stored in the system during an earthquake is approximately given by

$$\frac{1}{2}ku^2_{\max} = \frac{1}{2}kD^2 \tag{5.5}$$

And if there are no other processes dissipating energy, such as damping, and so on (although not entirely correct, we assume damping is negligible because it is small), then all the strain energy is transferred into kinetic energy:

$$\frac{1}{2}m\dot{u}^2_{\max} = \frac{1}{2}mS^2_v \tag{5.6}$$

From the principle of conservation of energy, these energy quantities must be equal; that is,

$$\frac{1}{2}kD^2 = \frac{1}{2}mS^2_v \tag{5.7}$$

This results in the definition of the spectral *pseudo-velocity, V,* in terms of the spectral displacement, *D,* as

$$\sqrt{\frac{k}{m}}D = V \Rightarrow \omega_n D = V \tag{5.8}$$

Note that in general $V \neq \dot{u}_{\max}$, but it is sufficiently close for most practical purposes. Figure 5.6 shows the pseudo-velocity and actual velocity response spectra due to El Centro with 5% damping, though the agreement between the response spectra is not as accurate as the comparison between acceleration response spectra.

From Equations 5.4 and 5.8 it can be seen that

$$\omega_n D = V = \frac{A}{\omega_n} \tag{5.9}$$

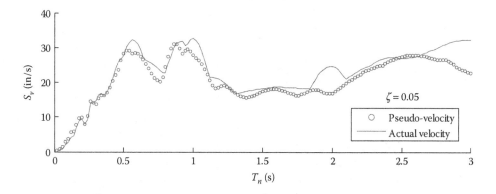

FIGURE 5.6   Comparison of actual and pseudo-velocity for El Centro quake.

or,

$$\frac{2\pi}{T_n} D = V = \frac{T_n}{2\pi} A \tag{5.10}$$

which means that D, V, and A contain the same information; thus, we can combine the three quantities into a single three-plot (tripartite) graph as a function of natural period, T_n, or natural frequency, ω_n.

Consider separately the relationships between D and V and that between V and A in Equation 5.9, and now take the logarithms of both sides of each equation,

$$\log V = \log \omega_n + \log D \tag{5.11}$$

$$\log V = -\log \omega_n + \log A \tag{5.12}$$

These represent equations of straight lines: the displacement with a slope of +1 and the acceleration with a slope of –1. That is, lines of constant D and A are inclined at +45° and –45°, respectively. Thus, to construct a tripartite graph, we need to plot V versus ω_n on vertical and horizontal logarithmic scales and add logarithmic scales inclined at +45° and –45° with respect to the vertical axis. So, V is read on the vertical scale, D is read on the scale at –45°, and A is read on the scale at +45°. When the argument is T_n, D is read from the scale at +45° and A is read from the scale at –45° (see Figure 5.7).

EXAMPLE 5.3

Given the structure shown in Figure E5.5, determine the response (D, V, and A) for the El Centro ground motion using the tripartite spectra in Figure 5.7.

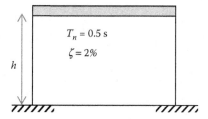

FIGURE E5.5   Schematic of frame structure in Example 5.3.

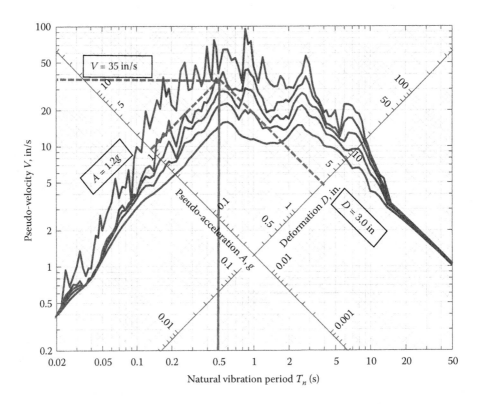

FIGURE 5.7 Response spectra for El Centro quake for $\zeta = 0\%$, 2%, 5%, 10%, and 20%, top to bottom. (Reproduced from Chopra, A. K., *Dynamics of Structures: Theory and Applications to Earthquake Engineering*, 4th edition, Prentice-Hall, Upper Saddle River, NJ, 2012. By permission of Prentice-Hall, Upper Saddle River, NJ.)

SOLUTION

Enter the 2%-damping response spectrum in Figure 5.7 with natural period of 0.5 s, and read $V = 35$ in/s from the vertical scale, $D = 7.5$ in from the scale at $+45°$, and $A = 1.2g$ from the scale at $-45°$. The lines showing these values are depicted in Figure 5.7.

EXAMPLE 5.4

The reinforced concrete bridge structure shown in Figure E5.6 is subjected to the El Centro earthquake; determine the total stiffness, structural period, deflection of the deck, and base shear. The deck weighs 750 kips and the substructure has 2% damping, modulus of elasticity, $E = 3000$ ksi, and the other geometric properties shown in Figure E5.6.

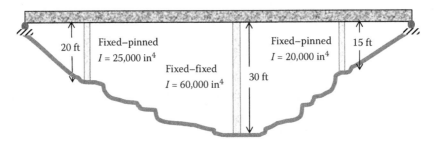

FIGURE E5.6 Schematic of reinforced concrete bridge structure.

SOLUTION

1. Mass, stiffness, and natural period of the system: The bridge can be modeled as an SDOF system assuming a rigid deck and only lateral deformations of the columns occur. The stiffness of the system is then the sum of column lateral stiffnesses. The mass and stiffness of the SDOF system are calculated as follows:
   Mass,

$$m = \frac{w}{g} = \frac{750\,\text{kips}}{386.4\,\text{in/s}^2} = 1.94\,\text{kips s}^2/\text{in}$$

Stiffness (see Table 3.1), the 30-foot column is fixed–fixed, while the other two are fixed–pinned,

$$k = \frac{3EI_1}{h_1^3} + \frac{12EI_2}{h_2^3} + \frac{3EI_3}{h_3^3}$$

$$= 3000\,\text{ksi}\left[\frac{3(25,000\,\text{in}^4)}{(20\,\text{ft}\times12\,\text{in/ft})^3} + \frac{12(60,000\,\text{in}^4)}{(30\,\text{ft}\times12\,\text{in/ft})^3} + \frac{3(20,000\,\text{in}^4)}{(15\,\text{ft}\times12\,\text{in/ft})^3}\right] = 93.4\,\text{kips/in}$$

The natural period of the system is

$$T_n = 2\pi\sqrt{\frac{m}{k}} = 2\pi\sqrt{\frac{1.94\,\text{kips s}^2}{93.4\,\text{kips/in}}} = 0.906\,\text{s}$$

2. Determine the maximum displacement: The maximum displacement can be determined using the response spectra depicted in Figure 5.7. Entering the 2%-damping response spectrum with $T_n = 0.91$ s, we can read the deck displacement from the scale at +45°,

$$D \cong 6\,\text{in}$$

3. Determine the maximum base shear: Calculate the maximum base shear at the support using the force–displacement relationship or equilibrium as described in Figure 3.24:

$$V_{bmax} = kD = 93.4\,\text{kips/in}\,(6\,\text{in}) = 565\,\text{kips}$$

A careful examination of the response spectra shown in Figure 5.7 shows that there are regions where each of the response parameters (D, V, and A) are nearly constant over a wide range of values for T_n. For example, for very flexible systems (those with large periods), the displacement is equal to the peak ground displacement for all damping ratios and experience negligible accelerations. However, stiff systems (those with small periods) accelerate at the PGA for all damping ratios and experience negligible relative displacements (only ground displacement). Also, damping has the greatest effect over the center region. This analysis of Figure 5.7 is specific to a set of response spectra for a specific earthquake. In order to compare the results of different earthquakes, we normalize each response parameter (spectral displacement, velocity, and acceleration) with respect to its corresponding peak ground value (peak ground spectral displacement, peak ground velocity, and PGA, respectively) as shown in Figure 5.8.

Figure 5.8 also depicts an idealized response spectrum, shown by a dashed line. We can observe that for the first portion ($T_n < 0.5$ s) of the response spectra, the acceleration is largest (acceleration sensitive), before decreasing precipitously; for the middle portion ($0.5\,\text{s} < T_n < 3\,\text{s}$), the velocity is largest (velocity sensitive); and for the last portion ($T_n > 3$ s), the displacement is largest (displacement sensitive). This phenomenon is one of the reasons for determining V and A. Furthermore, as

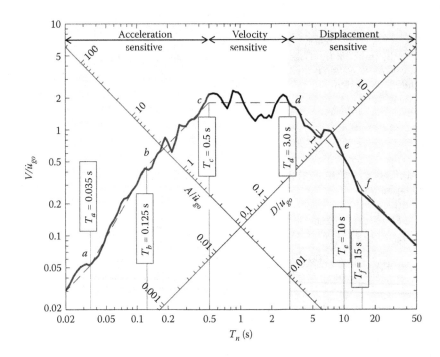

FIGURE 5.8 Spectral regions for El Centro quake; $\zeta = 5\%$. (Reproduced from Chopra, A. K., *Dynamics of Structures: Theory and Applications to Earthquake Engineering*, 4th edition, Prentice-Hall, Upper Saddle River, NJ, 2012. By permission of Prentice-Hall, Upper Saddle River, NJ.)

mentioned previously, delineating regions along the spectra where the different response quantities are dominant (or maximum) is useful in establishing design spectra, as discussed in Section 5.1.1.

5.1.1 ELASTIC DESIGN RESPONSE SPECTRUM

As discussed in Chapter 2, seismic ground accelerations vary greatly in magnitude of their peak amplitude and frequency content, both in space and time. For instance, two earthquake events of the same magnitude that strike a site at two different times will result in two different ground acceleration time histories even if the sites are at equal distances from the epicenter. For sites at different distances, seismic wave attenuation typically causes the seismic energy to diminish (in some cases, however, local soil conditions can amplify the effects for certain frequencies as was the case for the 1985 Mexico City earthquake discussed in Chapter 2). This makes it challenging to design for future buildings relying on past earthquake records to estimate the probable earthquake response. Therefore, we must assess the response of future structures to seismic loadings based on the probable PGAs (PPGA), with only secondary considerations to the frequency content. The PPGA can be determined from statistical analyses of several similar seismic events at sites with similar geological conditions.

An effective strategy to estimating accurate PPGA is to take averages of past earthquakes with similar characteristics, such as soil condition, epicentral distance, magnitude, and source mechanism. The statistical analysis uses normalized response spectra, assuming a Gaussian distribution for each point along the period axis as shown in Figure 5.9. The general procedure entails the following steps:

1. Select N earthquakes (10 for Figure 5.9) with similar characteristics.
2. Obtain D, V, and A for different T_n's for each earthquake; and normalize with respect to average peak ground displacement, u_{go}, peak ground velocity, $\dot{u}_{go}$, and PGA, $\ddot{u}_{go}$, respectively.

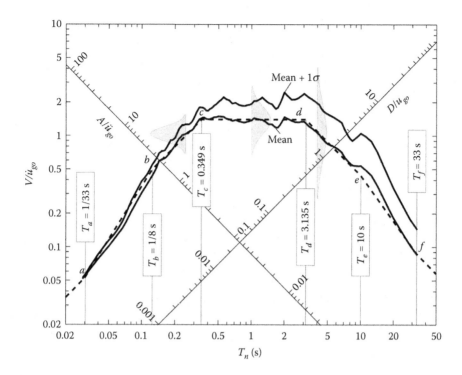

FIGURE 5.9 Mean and mean + 1σ spectra for a group of 10 quakes; ζ = 5%. (Reproduced from Chopra, A. K., *Dynamics of Structures: Theory and Applications to Earthquake Engineering*, 4th edition, Prentice-Hall, Upper Saddle River, NJ, 2012. By permission of Prentice-Hall, Upper Saddle River, NJ.)

3. Determine the mean and standard deviation (σ) at each period. Figure 5.9 schematically shows the probability distributions for V at $T_n = 0.25$, 1, and 4 s, which also show that the coefficient of variation (COV = σ/mean) changes with T_n.

Figure 5.9 also clearly shows that the graphs of the mean and the mean + 1σ are smoother than any individual parent spectrum, one of which is Figure 5.7.

The elastic design spectrum shown in Figure 5.10 is then obtained by smoothing out into a series of straight lines the spectra graphs. The dashed lines in Figures 5.9 and 5.10 represent the means of the response spectra. This simplified empirical approach to constructing elastic response spectra was first developed by Professors Nathan Newmark and William Hall, and is known as the Newmark–Hall method. The key parameters are the structural amplification factors, α, which represent the spectral amplification above ground motion. A summary of structural amplification factors is listed in Table 5.1. These empirical values were computed using records from firm ground, such as rock, soft rock, and competent sediments. Though limited in its applicability because of the small number of records and narrow site characteristics, many pieces of the general shape can be found in current code spectra, as will be demonstrated in Chapter 8.

After selecting a damping ratio, ζ, and a mean or mean + 1σ plot, the general procedure for drawing a design response spectrum entails the following steps:

1. Mark the key periods delineating the transition regions, T_a, T_b, T_e, and T_f. The periods delineating the velocity, acceleration, and displacement-controlled regions are obtained from the intersection of A and V and the intersection of V and D.
2. Plot lines corresponding to values of u_{go}, $\dot{u}_{go}$, and $\ddot{u}_{go}$ for the specified design ground motion. This peak ground motion graph represents the backbone of the response spectrum and can be considered the lower bound of the response.

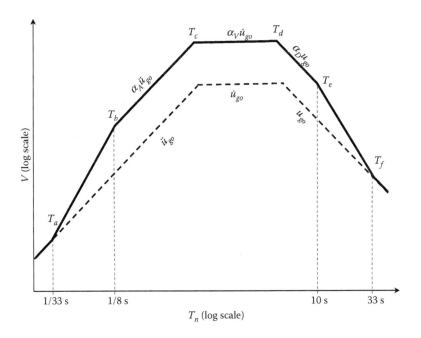

FIGURE 5.10 Newmark–Hall elastic design response spectrum.

TABLE 5.1
Structural Amplification Factors for Elastic Design Spectra

| Damping (%) | Mean (50th Percentile) | | | σ (84.1th Percentile) | | |
	α_A	α_V	α_D	α_A	α_V	α_D
2	2.74	2.03	1.63	3.66	2.92	2.42
5	2.12	1.65	1.39	2.71	2.30	2.01
10	1.64	1.37	1.20	1.99	1.84	1.69
ζ	$3.21-0.68\ln\zeta$	$2.31-0.41\ln\zeta$	$1.82-0.27\ln\zeta$	$4.38-1.04\ln\zeta$	$3.38-0.67\ln\zeta$	$2.73-0.45\ln\zeta$

Source: Newmark, N. M. and W.J. Hall, *Earthquake Spectra and Design*, Earthquake Engineering Research Institute, Oakland, CA, 1982.

3. Obtain values of α_A, α_V, and α_D for a specified damping ratio, ζ, from Table 5.1; where the mean or 50th percentile represents a 50% chance of being exceeded (50% exceedance).
4. Amplify the ground peak values (backbone) using these parameters to get: $\alpha_A\ddot{u}_{go}$, $\alpha_V\dot{u}_{go}$, and $\alpha_D u_{go}$.
5. Finally, connect the transition lines from *a* to *b* and from *e* to *f*.

The axes of the Newmark–Hall elastic design spectrum were normalized with respect to $\ddot{u}_{go} = 1g$, $\dot{u}_{go} = 48$ in/s, and $u_{go} = 36$ in. Values of $\ddot{u}_{go}$ can be obtained from deterministic or probabilistic site hazard analysis, whereas u_{go} and $\dot{u}_{go}$ can be determined from site hazard analyses or as empirical functions of $\ddot{u}_{go}$ when only $\ddot{u}_{go}$ is available. The following empirical relationships are recommended for estimating $\dot{u}_{go}$ and u_{go} in this case:

$$\frac{\dot{u}_{go}}{\ddot{u}_{go}} = 48g \text{ in/s} \quad \text{and} \quad \frac{\ddot{u}_{go} \cdot u_{go}}{\dot{u}_{go}^2} = 6 \tag{5.13}$$

A number of other researchers have conducted statistical analyses and determined values that vary considerably because these values depend on magnitude, epicentral distance, and soil characteristics at the recording station site.

EXAMPLE 5.5

Draw the 84.1th percentile response spectrum for 5% damping using the tripartite graph shown in Figure 5.11. The ground motion is anticipated to have a PGA of $\ddot{u}_{go} = 0.5g$.

SOLUTION

1. Determine the peak ground velocity, $\dot{u}_{go}$, and peak ground displacement, u_{go}: Use the PGA and Equation 5.13 to determine these quantities.
   Peak ground velocity,

$$\dot{u}_{go} = \left(48\ \frac{in/s}{g}\right)\ddot{u}_{go} = \left(48\ \frac{in/s}{g}\right)\cdot 0.5\,g = 24\ in/s$$

Peak ground displacement,

$$u_{go} = 6\dot{u}_{go}^2/\ddot{u}_{go} = 6(24\ in/s)^2/0.5(386\ in/s^2) = 18\ in$$

Plot the backbone quantities on the tripartite graph paper provided in Figure 5.11, as shown in Figure E5.7.

2. Determine the structural amplification factors, α_A, α_V, and α_D: For 84.1th percentile spectrum and 5% damping, these quantities are obtained from Table 5.1 as

$$\alpha_A = 2.71, \quad \alpha_V = 2.30, \quad \text{and} \quad \alpha_D = 2.01$$

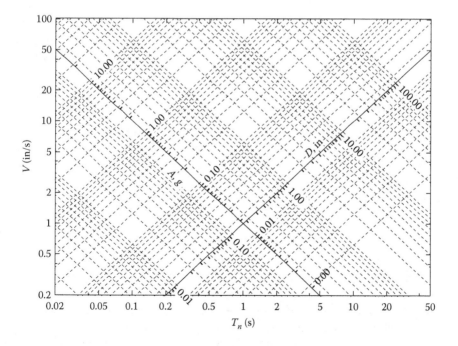

FIGURE 5.11 Tripartite graph paper.

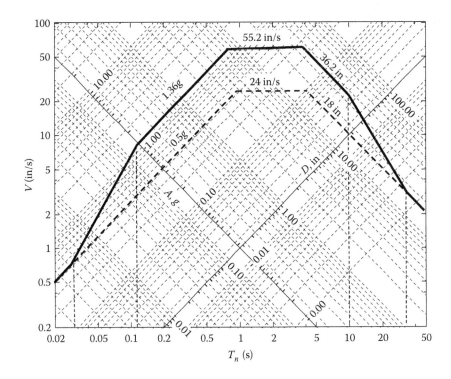

FIGURE E5.7   Elastic design response spectrum for peak ground acceleration of 0.5g and 5% damping.

3. Determine spectral quantities, A, V, and D: Amplify the backbone quantities to obtain A, V, and D.

$A = \ddot{u}_{go}\alpha_A = 0.5g\ (2.71) = 1.36g$
$V = \dot{u}_{go}\alpha_V = 24\ \text{in/s}\ (2.3) = 55.2\ \text{in/s}$
$D = u_{go}\alpha_D = 18\ \text{in}\ (2.01) = 36.2\ \text{in}$

   Plot these quantities in the same graph as the backbone quantities, Figure E5.7.

4. Complete the response spectrum by connecting the transition lines. Alternatively, the spectral graph can be obtained using a conventional graphing computer program such as MATLAB.

Note that the results of this example would be the same if we had derived the response spectrum for the standard peak ground values of $\ddot{u}_{go} = 1g$, $\dot{u}_{go} = 48\ \text{in/s}$, and $u_{go} = 36\ \text{in}$ and scaled the resulting spectrum by a factor of 0.5. Therefore, constructing an elastic design response spectrum using the Newmark–Hall method only requires the damping ratio and expected PGA, $\ddot{u}_{go} = \eta g$; The scaling factor, η, can easily be applied to the ordinates of the design spectra.

We have used the tripartite graph to delineate the three main regions of elastic design spectra (acceleration, velocity, and displacement sensitive); however, for design purposes, it is more convenient to return to the single acceleration spectra. This can be accomplished by plotting Equation 5.10 as shown in Figure 5.12 for the Newmark–Hall spectrum, $\ddot{u}_{go} = 1g$, $\dot{u}_{go} = 48\ \text{in/s}$, and $u_{go} = 36\ \text{in}$ and $\zeta = 5\%$. The axes of this spectrum are also plotted using linear scales in Figure 5.13. This figure is similar in shape to contemporary code-based design spectra.

In cases where only a single value of the pseudo-acceleration, A, is needed, it may not be necessary to construct the entire elastic design response spectrum; particularly if the period is between 1/8 and 10 s (constant acceleration, velocity, and displacement regions). For such a case, the governing design acceleration is given by the smallest of the following quantities:

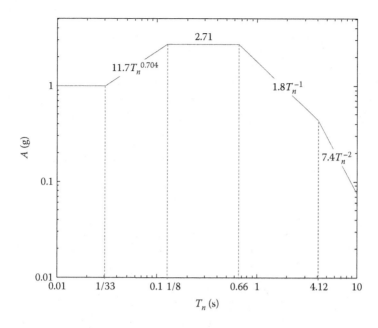

FIGURE 5.12 84.1th percentile acceleration response spectrum for $\zeta = 5\%$, plotted using logarithmic scales.

$$A = \ddot{u}_{go}\alpha_A$$
$$A = \omega_n V = 2\pi \dot{u}_{go}\alpha_V/T_n$$
$$A = \omega_n^2 D = (2\pi/T_n)^2 u_{go}\alpha_D$$

Also, as noted earlier, spectral quantities not only depend on PGA magnitude, but also on soil characteristics at the recording station site and epicentral distance. The values for α_A, α_V, and α_D in the Newmark–Hall method were derived based on a limited number of records that were obtained on a firm ground. More comprehensive studies have revealed that local site soil conditions have

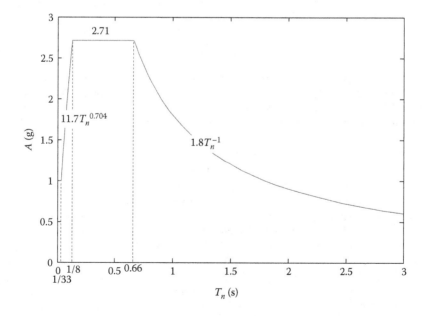

FIGURE 5.13 84.1th percentile acceleration response spectrum for $\zeta = 5\%$, plotted using linear scales.

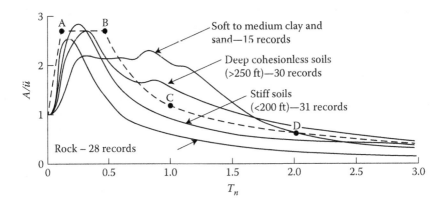

FIGURE 5.14 Influence of local soil conditions on response spectra for 5% damping. (Based in part on Seed, H. B. and I. M. Idriss, *Ground Motions and Soil Liquefactions during Earthquakes*, Earthquake Engineering Research Institute, Oakland, CA, 1982.)

a significant impact on spectra shapes. As shown in Figure 5.14, for soft soils, $\ddot{u}_{go}$ remains the same or decreases relative to firm soil, but $\dot{u}_{go}$ and u_{go} increase. Also, as the epicentral distance increases, the geology over the distance tends to filter the acceleration effects of certain frequencies. This results in the short-period portion of the design spectra being controlled by nearby earthquakes and the long-period portion of the design spectra controlled by earthquakes at further distances from the structure.

Contemporary design spectral analysis is based on a uniform hazard spectrum that accounts for the effects of all earthquakes expected at a site over a period of time using probability-based relations known as *attenuation relations*. These account for PGA magnitude, local site soil conditions, and epicentral distance all together; and along with probabilistic models of earthquake occurrence, these relations are used to obtain site-specific design spectra. Usually, however, the entire spectra are not computed following this procedure; rather, only accelerations corresponding to a couple of natural periods (0.2 and 1 s) are determined, which are then used to generate full design spectra using approximate relations. This approach will be discussed further in Chapter 8.

EXAMPLE 5.6

For the concrete bridge structure of Example 5.4, determine the deflection of the deck and base shear when the bridge is subjected to ground acceleration due to an earthquake characterized by the 84.1% design spectrum scaled to 0.32g PGA. The deck weighs 750 kips and the substructure has 5% damping, modulus of elasticity, $E = 3000$ ksi, and the other geometric properties are given in Example 5.4.

Solution

1. Mass, stiffness, and natural period of the system: The bridge can be modeled as an SDOF system assuming a rigid deck and considering only lateral deformations of the columns. The stiffness of the system is then the sum of column lateral stiffnesses. The mass and stiffness of the SDOF system are calculated in Example 5.4; the resulting quantities are

$$m = 1.94 \text{ kips s}^2/\text{in}$$

$$k = 93.4 \text{ kips/in}$$

$$T_n = 0.906 \text{ s}$$

The natural frequency of the system is

$$\omega_n = \sqrt{\frac{k}{m}} = \sqrt{\frac{93.4\,\text{kips/in}}{1.94\,\text{kips s}^2/\text{in}}} = 6.94\,\text{rad/s}$$

2. Determine the maximum displacement, D: To determine D due to an 84.1% design spectrum, scaled to 0.32g PGA, first, enter the response spectrum in Figure 5.12 with $T_n = 0.91$ s > 0.66 s, which is in the constant velocity region, so

$$V = \eta \cdot \dot{u}_{go}\alpha_V = 0.32(48\,\text{in/s})(2.3) = 35.3\,\text{in/s}$$

The maximum spectral displacement is

$$D = \frac{V}{\omega_n} = \frac{35.3\,\text{in/s}}{18.38\,\text{rad/s}} = 5.1\,\text{in}$$

3. Determine the maximum base shear: Calculate maximum base shear at the support using the force–displacement relationship or equilibrium as described in Figure 3.24:

$$V_{b\text{max}} = kD = 93.4\,\text{kips/in}(5.1\,\text{in}) = 476\,\text{kips}$$

EXAMPLE 5.7

The building frame shown in Figure E5.8, is subjected to a ground acceleration from an earthquake characterized by the 84.1% design spectrum scaled to 0.25g PGA, determine (i) maximum displacement u, (ii) the maximum base shear, and (iii) the maximum normal stresses in the columns ($I_x = 75$ in^4 and $S_x = 18$ in^3 for W8x21). Let us assume that the beam is rigid and system damping is 5%.

SOLUTION

1. Determine mass, stiffness, and natural period and frequency of the SDOF system: Since we have assumed the beam to be rigid, the stiffness of the system is given by the sum of column lateral stiffnesses. The mass and stiffness of the SDOF system are calculated as follows:
   Mass,

$$m = \frac{W}{g} = \frac{15\,\text{kips}(1000\,\text{lb/kip})}{386.4\,\text{in/s}^2} = 38.86\,\text{lb s}^2/\text{in}$$

Stiffness (see Table 3.1), when both column are fixed–pinned,

$$k = \frac{3EI}{L^3} + \frac{3EI}{L^3} = 2\left(\frac{3(29,000,000\,\text{psi})(75\,\text{in}^4)}{(15\,\text{ft}\times12\,\text{in/ft})^3}\right) = 2237.6\,\text{lb/in}$$

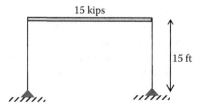

FIGURE E5.8   Schematic of frame structure in Example 5.7.

The natural period of the system is

$$T_n = 2\pi\sqrt{\frac{m}{k}} = 2\pi\sqrt{\frac{38.86 \text{ lb s}^2/\text{in}}{2237.6 \text{ lb/in}}} = 0.828 \text{ s}$$

The natural frequency of the system is

$$\omega_n = \sqrt{\frac{k}{m}} = \sqrt{\frac{2237.6 \text{ lb/in}}{38.86 \text{ lb s}^2/\text{in}}} = 7.59 \text{ rad/s}$$

2. Determine the maximum displacement, D: To determine D due to an 84.1% design spectrum, scaled to 0.25g PGA, first, enter the response spectrum in Figure 5.12 with $T_n = 0.83$ s > 0.66 s, so velocity is constant,

$$V = \eta \cdot \dot{u}_{go}\alpha_V = 0.25(48 \text{ in/ s})(2.3) = 27.6 \text{ in/s}$$

The maximum spectral displacement is

$$D = \frac{V}{\omega_n} = \frac{27.6 \text{ in/s}}{7.59 \text{ rad/s}} = 3.6 \text{ in}$$

3. Determine the maximum base shear: Calculate maximum base shear at the support for each column using the force–displacement relationship:

$$V_{max} = ku_o = \left(\frac{3EI}{L^3}\right)u_o = \left(\frac{3(29,000 \text{ ksi})(75 \text{ in}^4)}{(15 \text{ ft} \times 12 \text{ in/ft})^3}\right)3.6 \text{ in} = 4.1 \text{ kips}$$

4. Maximum normal stress due to bending: The maximum bending moment at the top of each column is calculated from equilibrium of the column member. Since the column has a pin support, summing moments about the top of the column will result in the following maximum internal bending moment (see Figure 3.24):

$$M_{max} = V_{max}L = 4.1 \text{ lb}(15 \text{ ft} \times 12 \text{ in/ft}) = 732 \text{ kip} \cdot \text{in}$$

Applying the flexure formula from mechanics of materials (Equation 3.60), the normal stress in the column due to the bending moment is determined as

$$\sigma_{max} = \frac{M_{max}}{S} = \frac{732 \text{ kip} \cdot \text{in}}{17 \text{ in}^3} = 43 \text{ ksi}$$

5.2 INELASTIC RESPONSE SPECTRUM

It is not practical or cost effective to design structures to behave in the elastic range for all possible seismic events. For the most severe cases, structural systems should be designed to resist seismic loads that cause strains beyond the elastic limit. Newmark and Hall also developed a procedure for accounting for inelastic behavior in a structural system response due to an earthquake—an inelastic response spectrum. To simplify the analysis, they assumed the behavior of the structure to be elastic-perfectly plastic (or elastoplastic) (see Figure 5.15). This assumption yields relatively accurate results and was also used to develop the inelastic design response spectrum. In this case, we solve the following nonlinear equation of motion:

$$m\ddot{u} + c\dot{u} + F_S(u) = -m\ddot{u}_g(t) \tag{5.14}$$

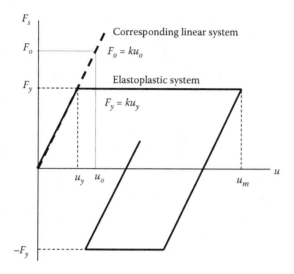

FIGURE 5.15   Elastoplastic and corresponding linear system behavior.

where $F_s(u)$ is the inelastic stiffness force.

All other parameters are the same as for elastic systems. The solution to this system requires numerical integration of the equation and the results can be presented in the same fashion as those for the elastic case, with one major difference, the ductility of the system, which is defined in terms of the ductility ratio μ. This parameter is defined as

$$\mu = \frac{u_m}{u_y} \Rightarrow u_m = \mu u_y \tag{5.15}$$

where:

u_m represents the maximum displacement of the SDOF system
u_y is the yield displacement of the SDOF system

We can also define a parameter based on the fictitious corresponding linear system and the yield force, known as the yield strength reduction factor, R_y:

$$R_y = \frac{F_o}{F_y} = \frac{ku_o}{ku_y} = \frac{u_o}{u_y} \tag{5.16}$$

where u_o represents the maximum displacement of the corresponding linear elastic system.

Note that when $u \leq u_y$, Equation 5.16 is related to a real elastic system, whereas for $u > u_y$, the system behaves inelastically and may not return to its equilibrium position after the shaking has stopped—it may experience permanent deformation.

In addition to μ, we can define a second displacement ratio between maximum displacement, u_m, and the maximum displacement of the fictitious corresponding linear system, u_o, which relates R_y and μ,

$$\frac{u_m}{u_o} = \frac{u_m}{u_o} \frac{u_y}{u_y} = \frac{\mu u_y}{u_o} = \frac{\mu}{R_y} \tag{5.17}$$

This gives the ductility demand imposed on an elastoplastic system by ground motion:

$$\mu = \frac{u_m}{u_o} R_y \tag{5.18}$$

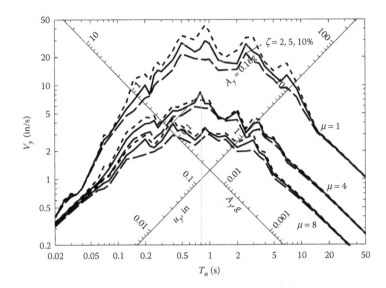

FIGURE 5.16 Elastoplastic response spectra for El Centro quake. (Reproduced from Chopra, A. K., *Dynamics of Structures: Theory and Applications to Earthquake Engineering*, 4th edition, Prentice-Hall, Upper Saddle River, NJ, 2012. By permission of Prentice-Hall, Upper Saddle River, NJ.)

For the purpose of designing a system, ductility capacity must be larger than the ductility demand, that is, the system must have the ability to deform beyond the elastic limit.

The comparison between the elastoplastic and the corresponding linear systems is also useful in establishing a procedure for using the elastic design response spectrum in Figure 5.10 to create the inelastic design response spectrum. Data presented by Chopra (2012) shows that in the displacement and velocity-sensitive regions, the maximum displacement is nearly equal to the ground displacement and the strength reduction factor is nearly equal to the ductility factor, that is, the result is independent of inelastic response. However, in the acceleration-sensitive region, the ductility demand is much greater than the strength reduction factor; short-period systems are stiffer and tend to attract more load, requiring a larger stiffness. Thus, we now have the response quantities (yield deformation, pseudo-velocity, and pseudo-acceleration) being functions of three different variables (period, damping ratio, and ductility ratio) (see Figure 5.16). This figure clearly indicates that damping has a decreasing effect as ductility increases and is much less effective than ductility in reducing the maximum response. Furthermore, to increase a system's earthquake resistance, it can be designed by either making it strong or ductile (or a combination of the two), because increasing ductility greatly decreases response.

Given yield displacement, u_y, the relationships between elastic (Equations 5.9) and inelastic quantities are as follows:

$$D_y = u_y$$
$$V_y = \omega_n D_y = \omega_n u_y$$
$$A_y = \omega_n^2 D_y = \omega_n^2 u_y = (2\pi/T_n)^2 D_y$$

And since $u_m = \mu u_y$ or $u_m = \mu D_y$,

$$u_m = \mu \left(\frac{T_n}{2\pi}\right)^2 A_y \tag{5.19}$$

EXAMPLE 5.8

Consider the building frame and characteristics given in Example 5.7. If the building frame is subjected to the El Centro earthquake, determine the maximum required column strength for a ductility ratio of 4.

SOLUTION

1. Mass, stiffness, and natural period and frequency of the SDOF system were determined in Example 5.7. That is, $m = 38.86$ lb s²/in, $k = 2237.6$ lb/in, $T_n = 0.828$ s, and $\omega_n = 7.59$ rad/s.

2. Determine the required column strength for a ductility demand of 4, F_y: To determine F_y, first, enter the response spectrum in Figure 5.16 with $T_n = 0.83$ s and read the pseudo-acceleration, A_y,

$$A_y \cong 0.16g$$

With the yield displacement, we obtain F_y,

$$F_y = mA_y = (38.86 \, (\text{lb} - \text{s}^2)/\text{in})(0.16)(386.4 \, \text{in/s}^2) = 2400 \, \text{lbs}$$

5.2.1 INELASTIC DESIGN RESPONSE SPECTRUM

The elastic design response spectrum is constructed using a backbone of peak ground properties and adjusting these values with the structural amplification factors. In a similar fashion, we can construct the inelastic design response spectrum using a new backbone of elastic spectral response properties and a new set of adjustment parameters that account for inelastic behavior. These adjustment parameters are functions of system ductility. Data analyses for elastic and inelastic constant spectral displacement and velocity regions show that the maximum inelastic displacement of the elastoplastic system is approximately equal to the maximum displacement of a linear system with the same period (see left-hand-side graph of Figure 5.17). First, we write the ductility ratio (Equation 5.15) in terms of forces using the system stiffness. That is,

$$\mu = \frac{u_m}{u_y} = \frac{kF_e}{kF_y} = \frac{F_e}{F_y} \tag{5.20}$$

Now we can rewrite this relationship in terms of accelerations using the system mass,

$$\mu = \frac{F_e}{F_y} = \frac{F_e}{F_i} = \frac{m\ddot{u}_y}{m\ddot{u}_i} = \frac{\ddot{u}_y}{\ddot{u}_i} \tag{5.21}$$

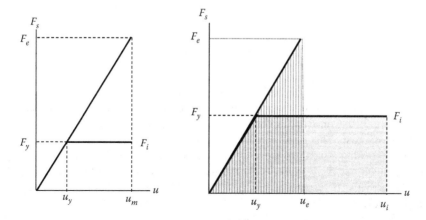

FIGURE 5.17 Force deformation curves for elastic and elastoplastic systems.

Since $F_i = F_y$. This results in the maximum inelastic acceleration being equal to the maximum elastic acceleration divided by the ductility factor,

$$\ddot{u}_i = \frac{\ddot{u}_y}{\mu} \Rightarrow A_y = \frac{A}{\mu} \tag{5.22}$$

The result presented in Equation 5.22 agrees relatively well with experimental data for the constant displacement and velocity regions, but not for the constant acceleration region. For the constant acceleration region, the relationship is obtained using the equivalence of energy in the elastic ($AREA_e$) and inelastic ($AREA_i$) systems. $AREA_e$ is the hatched area, while $AREA_i$ is the solid area in the right-hand-side graph of Figure 5.17. That is,

$$AREA_e = \frac{1}{2}u_e F_e = \frac{1}{2}F_y \frac{u_e^2}{u_y} \tag{5.23}$$

Since $k = F_e/u_e = F_y/u_y$ or $F_e = F_y u_e/u_y$,

$$AREA_i = \frac{1}{2}u_y F_y + (u_i - u_y)F_y \tag{5.24}$$

Setting Equations 5.23 and 5.24 equal, we obtain

$$\frac{1}{2}F_y \frac{u_e^2}{u_y} = \frac{1}{2}u_y F_y + (u_i - u_y)F_y \tag{5.25}$$

Using the definition of the ductility ratio, $\mu = u_i/u_y$, we can now solve for the inelastic displacement in terms of the elastic displacement,

$$u_i = \frac{\mu}{\sqrt{2\mu - 1}}u_e \Rightarrow D_y = \frac{\mu}{\sqrt{2\mu - 1}}D \tag{5.26}$$

In terms of the accelerations,

$$\ddot{u}_i = \frac{\ddot{u}_e}{\sqrt{2\mu - 1}} \Rightarrow A_y = \frac{A}{\sqrt{2\mu - 1}} \tag{5.27}$$

since $F_e = F_y u_e/u_y$, $F_y = m\ddot{u}_y = m\ddot{u}_i$, and $F_e = m\ddot{u}_e$.

With the adjustment parameters, we can easily construct the inelastic design response spectrum as shown in Figure 5.18. Constructing the inelastic design response spectrum follows similar procedures to the elastic case. That is, after identifying the ductility demand, μ, the value of the damping ratio, ζ, and choosing a mean or mean + 1σ plot, we can follow the details depicted in Figure 5.18, where the quantities A_y, V_y, and D_y are the inelastic acceleration, velocity, and displacement, respectively. Note that stiff systems having periods <1/33 s do not experience any adjustment—as it is for the elastic case.

As shown in the last section, elastic response spectra are typically given in terms of acceleration, which is how contemporary codes express elastic and inelastic seismic response.

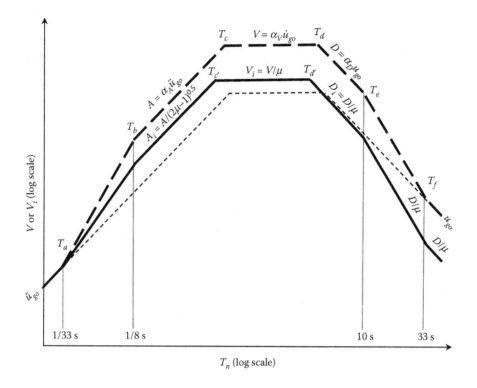

FIGURE 5.18 Newmark–Hall inelastic design response spectrum.

Thus, to obtain the maximum displacement, we use the relationship between displacement and pseudo-acceleration:

$$u_o = \mu \left(\frac{T_n}{2\pi} \right)^2 A_y \tag{5.28}$$

Also, the maximum force (required yield strength) can be determined using the relations between mass and acceleration:

$$F_o = mA_y \tag{5.29}$$

EXAMPLE 5.9

Consider the building frame shown in Figure E5.9 with $\zeta = 5\%$ and $I_x = 75$ in⁴ for the two columns subjected to a ground acceleration due to an earthquake characterized by the 84.1% design spectrum scaled to 0.5g PGA. Assume elastoplastic behavior and determine the lateral deformation and force for which the frame should be designed if $\mu = 4$.

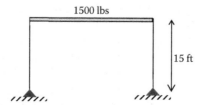

FIGURE E5.9 Building frame schematic.

SOLUTION

1. Mass, stiffness, and natural period of the SDOF system: Since we assumed the beam to be rigid, the stiffness of the system is given by the sum of column lateral stiffnesses. The mass and stiffness of the SDOF system are calculated as follows:
   Mass,

$$m = \frac{W}{g} = \frac{1500 \text{ lbs}}{386.4 \text{ in/s}^2} = 3.88 \text{ lb} \cdot \text{s}^2/\text{in}$$

   Stiffness (see Table 3.1), when both columns are fixed–pinned,

$$k = \frac{3EI}{L^3} + \frac{3EI}{L^3} = 2\left(\frac{3(29,000,000 \text{ psi})(75 \text{ in}^4)}{(15 \text{ ft} \times 12 \text{ in/ft})^3} \right) = 2237.6 \text{ lb/in}$$

   The natural period of the system is

$$T_n = 2\pi\sqrt{\frac{m}{k}} = 2\pi\sqrt{\frac{3.88 \text{ lb s}^2/\text{in}}{2237.6 \text{ lb/in}}} = 0.262 \text{ s}$$

2. Determine the maximum displacement for ductility of 4, u_o: First determine A due to an 84.1% design spectrum, scaled to 0.5g PGA, by entering the response spectrum in Figure 5.12 with $T_n = 0.26$ s, which is in the acceleration-sensitive region,

$$A = \eta \cdot 2.71g = 0.5(2.71g) = 1.335g$$

   The maximum displacement is computed using Equation 5.28 and Figure 5.18,

$$u_o = \mu\left(\frac{T_n}{2\pi}\right)^2 A_y = \mu\left(\frac{T_n}{2\pi}\right)^2 \frac{A}{\sqrt{2\mu - 1}} = 4\left(\frac{0.262 \text{ s}}{2\pi}\right)^2 \frac{1.335(386.4 \text{ in/s}^2)}{\sqrt{2(4) - 1}} = 1.353 \text{ in}$$

3. Determine the required column strength for a ductility of 4, F_y: To determine F_y, use $A = 1.335g$ and Equation 5.29.

$$F_y = mA_y = \frac{mA}{\sqrt{2\mu - 1}} = \frac{3.88 \text{ lb s}^2/\text{in}(1.355)(386.4 \text{ in/s}^2)A}{\sqrt{2(4) - 1}} = 770 \text{ lbs}$$

PROBLEMS

5.1 Use the MATLAB script provided in Chapter 4, Example 4.13 to draw response spectra for the displacement, velocity, and acceleration for an SDOF system subjected to the 1994 Northridge earthquake ground acceleration for 2%, 5%, and 10% damping.

5.2 The building frame shown is subjected to the El Centro earthquake; determine the total stiffness, structural period, deflection of the beam, base shear, and bending stresses in each of the columns. The roof weighs 10 kips and the substructure has 5% damping, modulus of elasticity, $E = 29,000$ ksi, and the other geometric properties shown.

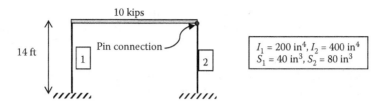

5.3 The following structural system is subjected to the El Centro earthquake; determine the total stiffness, structural period, deflection of the beam, and base shear. The substructure has 2% damping, modulus of elasticity, $E = 3000$ ksi, and the other geometric properties as shown.

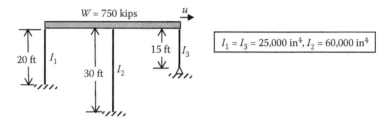

5.4 A 25 kip sensitive piece of equipment will be installed on the roof of the following frame shown with a rigid slab and four columns with $EI_x = 125,000$ kip in² and 5% damping. The frame is located in Mexico City at a site with response spectrum depicted in Example 5.2. Determine approximately the magnitude of the shear force ($V_{req'd}$) needed to design the equipment anchorage.

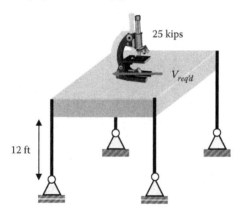

5.5 Use a tripartite graph to draw the elastic 50th percentile design response spectrum with a maximum ground acceleration of 0.25g and a damping ratio of 2%.

5.6 For the frame of Problem 5.2, determine the deflection of the beam, base shear, and bending stresses when the frame is subjected to ground acceleration due to an earthquake characterized by the 84.1% design spectrum scaled to 0.5g peak ground acceleration. The substructure has 5% damping and the other geometric properties given in Problem 5.2.

5.7 Burn's tower (a water tank weighing 4200 kips) on the campus of University of the Pacific is supported on a 120-ft high cantilever tower with a stiffness of 2000 kips/in and 5% damping. Determine the design values for lateral deformation and base shear if the tower is subjected to ground acceleration due to an earthquake characterized by the 84.1% design spectrum scaled to 0.3g peak ground acceleration.

5.8   Use a tripartite graph to draw the inelastic 50th percentile design response spectrum with a maximum ground acceleration of 0.25g, a damping ratio of 2%, and a ductility ratio of 5.

5.9   Given the following 14-ft. steel tower subjected to a ground acceleration due to an earthquake characterized by the 84.1% design spectrum scaled to 0.25g peak ground acceleration, determine (i) maximum lateral displacement, and (ii) the maximum base shear. Assume system damping of 5% and ductility of 5.

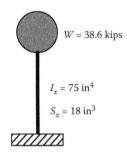

$W = 38.6$ kips

$I_x = 75$ in^4

$S_x = 18$ in^3

5.10 Consider the following building frame with a rigid beam that is pinned connected to one column and rigidly connected to the other as shown; each column has $EI = 40,000,000$ kip in^2. The frame is subjected to ground acceleration due to an earthquake characterized by the 84.1% design spectrum scaled to 0.33g peak ground acceleration. Determine maximum stress in each column given that $S_x = 2,000$ in^3 and damping is 5%.

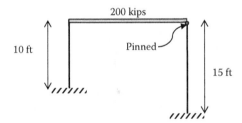

200 kips

10 ft

Pinned

15 ft

5.11 Solve for the maximum base shear in the frame shown in problem 5.10 assuming a ductility ratio of 4.

5.12 Consider a one-story frame with lumped weight w and natural vibration period, $T = 0.25$ s, in the elastic range. Determine the lateral deformation and force in terms of w for which the frame should be designed if $\mu = 6$. Assume that $\zeta = 5\%$, elastoplastic behavior, and an 84.1% design earthquake with $\ddot{u}_{go} = 0.25g$.

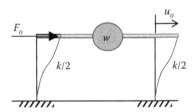

F_o

u_o

w

$k/2$ $k/2$

5.13 Two identical power transformers to be erected adjacent to each other are supported by steel ($E = 29,000$ ksi) pedestals and can be idealized as SDOF systems. The pedestals are designed to withstand inelastic deformations of up to six times their yield values. Using an inelastic 50th percentile design response spectrum with a maximum ground acceleration of 0.25g, estimate the minimum clear distance (Δ) that must be left between the two transformers to avoid collision. $I = 1380$ in^4 and $W = 1.5$ kips.

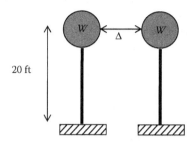

REFERENCES

Chopra, A. K., *Dynamics of Structures: Theory and Applications to Earthquake Engineering*, 4th edition, Prentice-Hall, Upper Saddle River, NJ, 2012.

Seed, H. B. and I. M. Idriss, *Ground Motions and Soil Liquefactions during Earthquakes*, Earthquake Engineering Research Institute, Oakland, CA, 1982.

6 Generalized SDOF System Analysis

After reading this chapter, you will be able to:

1. Setup the equation of motion for generalized SDOF systems
2. Determine the generalized properties, including natural period and participation factor for generalized shear building systems and distributed mass/elasticity systems
3. Use response spectra and participation factors to calculate the maximum base shear and overturning moment response of generalized shear building systems and distributed mass/elasticity systems

Before presenting the formulation for a system discretized into multiple degrees of freedom, it is instructive to study the development of an approach that condenses these MDOF systems into a SDOF case. Unlike the MDOF system formulation, this approach does not require simultaneous manipulation of multiple algebraic equations. The key to obtaining relatively accurate results with a generalized SDOF system analysis is selecting an appropriate shape function for the deformed system during vibration.

In this chapter, we also describe the analysis of systems with distributed mass and stiffness as generalized SDOF systems. While this analysis approach is not as precise as modeling a system with an infinite number of degrees of freedom, modeling systems as generalized SDOF systems is more accurate than approximating the entire system as a simple SDOF system as discussed in Chapter 3. Furthermore, utilizing an infinite number of degrees of freedom to model the distributed mass and stiffness as a function of position would require formulation in terms of partial differential equations in both position and time.

The generalized SDOF system analysis presented in this chapter is based on replacing the required partial differential equation in time and space with an approximate differential equation in time only. The space variation is tracked using an approximate equation of the mode of vibration for the system. The process entails selecting a point of interest along the distributed system and tracking its motion with respect to time using an SDOF analysis; in order to obtain the response of all other points, we scale the magnitude of the point in question based on the approximate equation of the mode of vibration. The solution to the differential equation representing the generalized SDOF system follows the standard procedures presented in Chapter 3; and we can apply all the solution techniques discussed thus far, including the response spectrum analysis presented in Chapter 5.

Replacing the distributed dynamic properties (mass, damping, and stiffness) with lumped properties at the point being tracked requires integrating the variation in these properties along the length of the system. For discrete MDOF systems, the process involves combining the discrete properties associated with each DOF into a set of generalized properties. These generalized properties are most useful in developing the participation factor, which is critical in applying the results from an SDOF system analysis to an MDOF system analysis.

The formulation of the equation of motion of a generalized SDOF system entails condensing all degrees of freedom into one. In the following sections, we present analysis procedures for a multistory shear building frame (discrete system) and a cantilever tower with distributed properties (continuous system), both subjected to earthquake ground motions. In each case, we assume axial deformations to be negligible during the dynamic excitation, which leaves the lateral translation (or deflection) as the only possible deformation of the system.

6.1 DISCRETE SYSTEM (SHEAR BUILDINGS)

Multistory building systems can be modeled as having a discrete number of degrees of freedom, the equation of motion for which can be formulated by assuming the building mass to be distributed as concentrated masses at various levels, as in the simplified lumped mass tower shown in Figure 6.1; see Section 3.1.2 for a discussion on how to determine the lumped mass at each level. The deformed shape of the system along with the virtual deformation used in the formulation of the equation of motion is also depicted in the figure. The actual deformation of this system is a function of position and time. The time effect is only tracked at the top of the building using $z(t)$, whereas the position of the entire shear building is followed using an assumed shape vector, ψ_j, which is typically determined by evaluating a shape function at each level. Also shown in Figure 6.1 is an equivalent generalized SDOF system.

6.1.1 EQUATION OF MOTION

The actual motion of the shear building shown in Figure 6.1 can be obtained by determining the motion of the top of the building, $z(t)$, and scaling it to the other levels using an appropriate shape vector, ψ_j. That is, at an arbitrary level, j, the relative displacement is

$$u_j(t) = \psi_j \cdot z(t) \qquad (6.1)$$

where:
$j = 1, 2, \ldots, N$

The motion at the top of the building, $z(t)$, can be obtained using a generalized equation of motion for the equivalent SDOF system shown in Figure 6.1. This equation of motion can be obtained using D'Alembert's principle as discussed in Chapter 3 or the principle of virtual work, which is the approach we follow in this chapter. The resulting equation is of the same form as the one

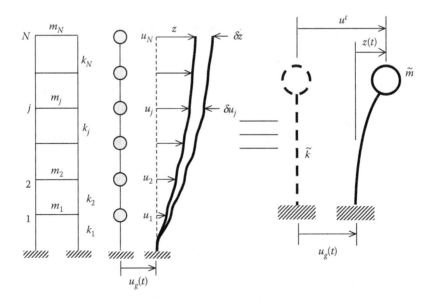

FIGURE 6.1 MDOF shear building and equivalent generalized SDOF system.

that characterizes the vibration of the SDOF oscillator introduced in Chapter 3; thus, all analysis procedures covered so far are applicable to this case.

The shape vector, ψ_j, can be obtained from any function, $\psi(x)$, provided it satisfies the geometric boundary conditions. However, the accuracy of the results depends on how close the assumed shape function approximates the actual shape of the deformed frame. Also, the function $\psi(x)$ must be normalized at the top of the building such that $\psi(H) = 1.0$, where H is the height of the building. Figure 6.2 presents several possible shape functions based on building elevation aspect ratio.

To account for the ground excitation loading in the formulation of the equation of motion, we also need the total displacement at level j, which is a combination of the relative displacement, $u_j(t)$, and the ground level lateral displacement, $u_g(t)$:

$$u_j^t(t) = u_j(t) + u_g(t) \tag{6.2}$$

We can now use the principle of virtual work to derive the generalized equation of motion. This principle states that the virtual work of all internal forces, δW_I, equals the virtual work of all external forces, δW_E, for a given virtual displacement, δu. That is,

$$\delta W_I = \delta W_E \tag{6.3}$$

where δW_I includes the contribution from stiffness and damping forces:

$$\delta W_I = \delta W_{\text{stiffness}} + \delta W_{\text{damping}} \tag{6.4}$$

And δW_E includes the contribution from inertial forces of each mass:

$$\delta W_E = \delta W_{\text{inertial}} \tag{6.5}$$

The virtual work of each of these groups of forces is given by summation of the product of the magnitude of the force and the corresponding virtual displacement. The stiffness and damping forces can be viewed as effective shear forces in each story, whereas the inertial forces can be regarded as effective lateral loads applied at each level. The stiffness forces are functions of *story*

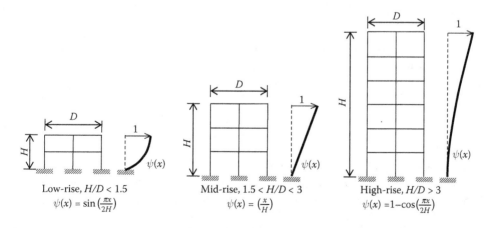

Low-rise, $H/D < 1.5$
$\psi(x) = \sin\left(\frac{\pi x}{2H}\right)$

Mid-rise, $1.5 < H/D < 3$
$\psi(x) = \left(\frac{x}{H}\right)$

High-rise, $H/D > 3$
$\psi(x) = 1 - \cos\left(\frac{\pi x}{2H}\right)$

FIGURE 6.2 Possible shape functions based on building elevation aspect ratio. (With kind permission from Springer Science+Business Media: Naeim, F. 2001. *The Seismic Design Handbook*, 2nd edition, Dordrecht, The Netherlands.)

drift (relative displacement from one floor to the next); the damping forces are functions of relative velocity from one floor to the next; and the inertial forces are functions of the total acceleration at each level.

First, determine the contribution of the stiffness forces to the virtual work equation. The stiffness force at level j, f_{sj}, is equal to the corresponding story stiffness times story drift,

$$f_{sj} = k_j(u_j - u_{j-1}) \tag{6.6}$$

where k_j is the sum of all the lateral stiffness systems (columns, shear walls, bracing, etc.) in the direction in question at story j. Substituting Equation 6.1 evaluated at stories j and j–1 into this equation yields

$$f_{sj} = k_j(z(t)\psi_j - z(t)\psi_{j-1}) \tag{6.7}$$

We can factor out $z(t)$ and replace the shape vector values with the change in values between two stories, $(\Delta\psi_j = \psi_j - \psi_{j-1})$,

$$f_{sj} = k_j z(t) \cdot \Delta\psi_j \tag{6.8}$$

The virtual work of this force is

$$\delta W_{\text{stiffness}\,j} = f_{sj}(\delta u_j - \delta u_{j-1}) \tag{6.9}$$

where δu_j represents the internal virtual displacement of the structure. Substituting the value of f_{sj} from Equation 6.8 and virtual values of Equation 6.1 into Equation 6.9 results in the internal virtual work due to stiffness,

$$\delta W_{\text{stiffness}\,j} = k_j z(t) \cdot \Delta\psi_j(\delta z\psi_j - \delta z\psi_{j-1}) \tag{6.10}$$

We can factor out δz and substitute $\Delta\psi_j = \psi_j - \psi_{j-1}$,

$$\delta W_{\text{stiffness}\,j} = \delta z \cdot z(t)k_j \cdot \Delta\psi_j^2 \tag{6.11}$$

We can then add the contribution of all N stories to obtain the total internal work from the stiffness forces in each story,

$$\delta W_{\text{stiffness}} = \delta z \cdot z(t)\sum_{j=1}^{N} k_j \cdot \Delta\psi_j^2 \tag{6.12}$$

Similarly, we can determine the contribution of the damping forces to the virtual work equation. The damping force at level j, f_{dj}, is equal to the corresponding story damping times story relative velocity,

$$f_{dj} = c_j(\dot{u}_j - \dot{u}_{j-1}) \tag{6.13}$$

where c_j is the sum of all the damping systems in the direction in question at story j.

Substituting Equation 6.1 evaluated at stories j and $j-1$ into this equation yields

$$f_{dj} = c_j(\dot{z}(t)\psi_j - \dot{z}(t)\psi_{j-1}^*)$$
(6.14)

We can factor out the velocity and substitute $\Delta\psi_j = \psi_j - \psi_{j-1}$,

$$f_{dj} = c_j\dot{z}(t) \cdot \Delta\psi_j$$
(6.15)

The virtual work of this force is

$$\delta W_{\text{damping}\,j} = f_{dj}(\delta u_j - \delta u_{j-1})$$
(6.16)

Substituting the value of f_{dj} from Equation 6.15 and virtual values of Equation 6.1 into Equation 6.16 results in the internal virtual work due to damping,

$$\delta W_{\text{damping}\,j} = c_j\dot{z}(t) \cdot \Delta\psi_j(\delta z\psi_j - \delta z\psi_{j-1})$$
(6.17)

We can factor out δz and substitute $\Delta\psi_j = \psi_j - \psi_{j-1}$,

$$\delta W_{\text{damping}\,j} = \delta z \cdot \dot{z}(t)c_j \cdot \Delta\psi_j^2$$
(6.18)

We can then add the contribution of all N stories to obtain the total internal work from the damping forces in each story,

$$\delta W_{\text{damping}} = \delta z \cdot \dot{z}(t)\sum_{j=1}^{N} c_j \cdot \Delta\psi_j^2$$
(6.19)

Substituting Equations 6.12 and 6.19 into Equation 6.4, we obtain the total internal virtual work,

$$\delta W_I = \delta z \cdot z(t)\sum_{j=1}^{N} k_j \cdot \Delta\psi_j^2 + \delta z \cdot \dot{z}(t)\sum_{j=1}^{N} c_j \cdot \Delta\psi_j^2$$
(6.20)

Finally, we determine the contribution of the external forces to the virtual work equation. The inertial force at level j, f_{Ij}, is equal to the corresponding mass times story total acceleration,

$$f_{Ij} = m_j\ddot{u}_j^t$$
(6.21)

Substituting the second derivative with respect to time of Equation 6.2 into this equation yields

$$f_{Ij} = m_j(\ddot{u}_g + \ddot{u}_j)$$
(6.22)

Now taking the second time derivative of Equation 6.1 evaluated at story j and substituting into Equation 6.22 gives,

$$f_{Ij} = m_j(\ddot{u}_g + \ddot{z}(t)\psi_j)$$
(6.23)

The external virtual work of the inertial force is

$$^\bullet\delta W_{\text{inertial } j} = -f_{Ij}\delta u_j \qquad (6.24)$$

This is negative because the sense of the force is opposite to the direction of the virtual displacement. Substituting the value of f_{Ij} from Equation 6.23 and virtual values of Equation 6.1 yields

$$\delta W_{\text{inertial } j} = -m_j(\ddot{u}_g + \ddot{z}(t)\psi_j)\delta z\psi_j \qquad (6.25)$$

We can now sum the contribution of all N stories to obtain the total external work from the inertial forces at each level j,

$$\delta W_E = -\delta z \cdot \ddot{u}_g \sum_{j=1}^{N} m_j\psi_j - \delta z \cdot \ddot{z}(t)\sum_{j=1}^{N} m_j\psi_j^2 \qquad (6.26)$$

We can now apply the principle of virtual work (Equation 6.3) by setting the internal virtual work, Equation 6.20, equal to the external virtual work, Equation 6.26,

$$\delta z \cdot z(t)\sum_{j=1}^{N} k_j \cdot \Delta\psi_j^2 + \delta z \cdot \dot{z}(t)\sum_{j=1}^{N} c_j \cdot \Delta\psi_j^2 = -\delta z \cdot \ddot{u}_g \sum_{j=1}^{N} m_j\psi_j - \delta z \cdot \ddot{z}(t)\sum_{j=1}^{N} m_j\psi_j^2 \qquad (6.27)$$

After eliminating the virtual displacement, δz, since it appears in each term, and rearranging this equation we obtain

$$z(t)\sum_{j=1}^{N} k_j \cdot \Delta\psi_j^2 + \dot{z}(t)\sum_{j=1}^{N} c_j \cdot \Delta\psi_j^2 + \ddot{z}(t)\sum_{j=1}^{N} m_j\psi_j^2 = -\ddot{u}_g \sum_{j=1}^{N} m_j\psi_j \qquad (6.28)$$

This is the equation of motion of a generalized SDOF system, which can be written in the same form as the equation of motion derived in Chapter 3,

$$\tilde{m}\ddot{z} + \tilde{c}\dot{z} + \tilde{k}z = -\tilde{L}\ddot{u}_g(t) \qquad (6.29)$$

where:

$\tilde{m} = \displaystyle\sum_{j=1}^{N} m_j \cdot \psi_j^2$ is the generalized mass

$\tilde{k} = \displaystyle\sum_{i=1}^{N} k_j \cdot (\Delta\psi_j)^2$ is the generalized stiffness

$\tilde{c} = \displaystyle\sum_{i=1}^{N} c_j \cdot (\Delta\psi_j)^2$ is the generalized damping

$\tilde{L} = \displaystyle\sum_{i=1}^{N} m_j \cdot \psi_j$ is the generalized force

N in the summation is the number of stories in the building
$\Delta\psi_j$ is the relative value of the shape function of two consecutive levels, that is, $\Delta\psi_j = \psi_j - \psi_{j-1}$
m_j is the mass at level j
k_j is the stiffness for the jth story
c_j is the damping for the jth story

6.1.2 GENERALIZED PARTICIPATION FACTOR, FREQUENCY, AND NATURAL PERIOD

Rearranging the generalized equation of motion (Equation 6.29) into the form of Equation 3.42, we obtain

$$\ddot{z} + 2\zeta\omega_n\dot{z} + \omega_n^2 z = -\tilde{\Gamma}\ddot{u}_g(t) \tag{6.30}$$

where:
the generalized natural frequency is given as

$$\omega_n = \sqrt{\frac{\tilde{k}}{\tilde{m}}} \tag{6.31}$$

the generalized participation factor is given as

$$\tilde{\Gamma} = \frac{\tilde{L}}{\tilde{m}} \tag{6.32}$$

ζ is an estimate of the damping ratio.

We can also establish the generalized natural period,

$$T_n = 2\pi\sqrt{\frac{\tilde{m}}{\tilde{k}}} \tag{6.33}$$

6.1.3 DEFLECTIONS, BASE SHEAR, AND MOMENTS USING PARTICIPATION FACTORS AND RESPONSE SPECTRA

After finding the period, T_n, using Equation 6.33, we can determine the maximum dynamic response caused by any dynamic excitation (or force) using the SDOF system analysis discussed in Chapters 3 and 4. Also, we can use the response spectrum analysis method discussed in Chapter 5 to determine the peak response to an earthquake characterized by a design response spectrum with spectral displacement, velocity, and acceleration (D, V, and A, respectively) for an appropriate damping ratio. For this case, the original ground acceleration, $u_g(t)$, is multiplied by the participation factor, $\tilde{\Gamma}$. Thus, the solution to the equation of motion based on response spectrum analysis using the spectral displacement, D, gives the maximum displacement at the top of the building, relative to the ground, as

$$z = \tilde{\Gamma} \cdot D \tag{6.34}$$

This maximum displacement can be distributed to the other floors using the shape vector, ψ_j (Equation 6.1), as shown in Figure 6.3,

$$u_{jo} = \psi_j \cdot z = \psi_j \cdot \tilde{\Gamma} \cdot D \tag{6.35}$$

This can also be written in terms of the spectral acceleration, A as,

$$u_{jo} = \psi_j \frac{\tilde{\Gamma}}{\omega_n^2} A \tag{6.36}$$

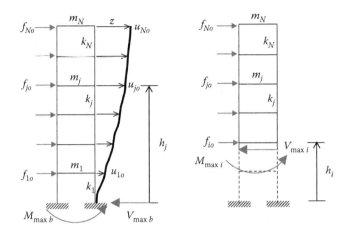

FIGURE 6.3 Maximum dynamic displacements and associated equivalent static forces.

The acceleration is proportional to the displacement; that is, $A = \omega_n^2 D$.

The equivalent static forces associated with these floor displacements can be obtained using the product of the maximum floor displacement and associated stiffness or the product of the associated story mass and maximum story acceleration; that is,

$$f_{jo} = \tilde{\Gamma} m_j \psi_j A \tag{6.37}$$

Also, as discussed in Section 3.3, with these forces we can conduct a static structural analysis to determine element forces (bending moment, shear force, and axial force) and stresses needed for design of structural elements; no additional dynamic analysis is necessary. Figure 6.3 shows these equivalent static forces along with the base shear and overturning moment. The internal story shear force, $V_{\max i}$, and internal story moment, $M_{\max i}$, at an arbitrary level i can be obtained by applying static equilibrium to the right-hand-side free-body diagram,

$$V_{\max i} = \sum_{j=i}^{N} f_{jo} \tag{6.38}$$

$$M_{\max i} = \sum_{j=i}^{N} (h_j - h_i) f_{jo} \tag{6.39}$$

Setting i equal to 1 in Equations 6.38 and 6.39 results in the base shear force, $V_{\max b}$, and overturning moment, $M_{\max b}$,

$$V_{\max b} = \sum_{j=1}^{N} f_{jo} \tag{6.40}$$

$$M_{\max b} = \sum_{j=1}^{N} h_j f_{oj} \tag{6.41}$$

In the derivation of Equations 6.39 through 6.41, it is assumed that the floor weights are directly in the center of the building frame; thus, making no contribution to the moment equilibrium equation. However, for buildings with off-center weights, a full moment static equilibrium analysis should be conducted to account for the contribution of these off-center forces.

EXAMPLE 6.1

Given the three-story building shown in Figure E6.1 frame with damping of 5% and subjected to a ground acceleration due to a 0.25g peak ground acceleration (PGA) earthquake characterized by an 84.1% design spectrum, determine (i) peak displacements, (ii) maximum base shear, and (iii) maximum floor overturning moments. Let us assume that beams are rigid. Each story has a stiffness $k = 326.3$ kips/in and a shape function appropriate for a mid-rise building is to be used.

SOLUTION

1. Determine the shape vector, ψ_j. This can be obtained using Figure 6.2 for a mid-rise building function, $\psi_j = x/H$, where $H = 36$ ft.

Level, j	x (ft)	ψ (x)	ψ_j
3	36	36/H	1
2	24	24/H	2/3
1	12	12/H	1/3

ψ_j in the vector form,

$$\psi_j = \begin{Bmatrix} 1/3 \\ 2/3 \\ 1 \end{Bmatrix}$$

2. Determine the generalized properties. The building frame can be modeled as a generalized SDOF system assuming that the beams are rigid and only lateral deformations of columns occur. The stiffness of each floor is then obtained by summing the column lateral stiffnesses at each level. The generalized mass, stiffness, and force of the generalized SDOF system are calculated as follows:

Level, j	m_j	k_j	ψ_j	$\Delta\psi_j$	$m_j \cdot \psi_j$	$m_j \cdot \psi_j^2$	$k_j \cdot (\Delta\psi_j)$
3	m/2	k	1	1/3	$m/2 \cdot (1)$	$m/2 \cdot (1)^2$	$k \cdot (1/3)^2$
2	m	k	2/3	1/3	$m \cdot (2/3)$	$m \cdot (2/3)^2$	$k \cdot (1/3)^2$
1	m	k	1/3	1/3	$m \cdot (1/3)$	$m \cdot (1/3)^2$	$k \cdot (1/3)^2$
$\sum_{j=1}^{3} =$					1.5 m	19/18 m	1/3 k

Substituting the given values for mass and stiffness, the following properties of the generalized SDOF are calculated:
Generalized mass,

$$\tilde{m} = \sum_{j=1}^{3} m_j \cdot \psi_j^2 = \frac{19}{18} m = \frac{19}{18} \frac{W}{g} = \frac{19}{18} \frac{100 \text{ kips}(1000 \text{ lb/kip})}{386.4 \text{ in/s}^2} = 273.2 \text{lb s}^2/\text{in}$$

Generalized stiffness,

$$\tilde{k} = \sum_{j=1}^{3} k_j \cdot (\Delta\psi_j)^2 = \frac{1}{3} k = \frac{1}{3} 326.3 \text{ kips/in}(1000 \text{ lb/kip}) = 108,767 \text{ lb/in}$$

Generalized force,

$$\tilde{L} = \sum_{j=1}^{3} m_j \cdot \psi_j = 1.5\,m = 1.5\frac{W}{g} = 1.5\frac{100\ \text{kips}(1000\ \text{lb/kip})}{386.4\ \text{in/s}^2} = 388.2\ \text{lb s}^2/\text{in}$$

3. Determine the natural period, frequency, and participation factor of the generalized SDOF system:
   Natural period,

$$T_n = 2\pi\sqrt{\frac{\tilde{m}}{\tilde{k}}} = 2\pi\sqrt{\frac{273.2\ \text{lb s}^2/\text{in}}{108{,}767\ \text{lb/in}}} = 0.315\,\text{s}$$

   Natural frequency,

$$\omega_n = \sqrt{\frac{\tilde{k}}{\tilde{m}}} = \sqrt{\frac{108{,}767\ \text{lb/in}}{273.2\ \text{lb s}^2/\text{in}}} = 19.95\,\text{rad/s}$$

   Participation factor,

$$\tilde{\Gamma} = \frac{\tilde{L}}{\tilde{m}} = \frac{388.2\ \text{lb s}^2/\text{in}}{273.2\ \text{lb s}^2/\text{in}} = 1.421$$

4. Determine maximum floor displacements: Determine the maximum displacement at the top of the building, $z(t)$ or D due to an 84.1% design spectrum, scaled to 0.25g PGA; first, enter the response spectrum in Figure 5.11 with $T_n = 0.315$ s, which is greater than 1/8 s so the acceleration is constant,

$$A = \eta \cdot 2.71g = 0.25(2.71)(386.4\ \text{in/s}^2) = 261.8\ \text{in/s}^2$$

where:
   η is the PGA scale for a given earthquake; in this case 0.25

The maximum spectral displacement is

$$D = \frac{A}{\omega_n^2} = \frac{261.8\ \text{in/s}^2}{(19.95\ \text{rad/s})^2} = 0.658\ \text{in}$$

We can now determine the maximum floor displacements using Equation 6.35,

$$u_{jo} = \psi_j\tilde{\Gamma}\cdot D = \begin{Bmatrix} 1/3 \\ 2/3 \\ 1 \end{Bmatrix}(1.421)(0.658\ \text{in}) = \begin{Bmatrix} 0.311 \\ 0.623 \\ 0.934 \end{Bmatrix}\text{in}$$

5. Determine the equivalent static story forces: Determine the equivalent static story forces using Equation 6.37,

$$f_{jo} = \tilde{\Gamma}m_j\psi_j A$$

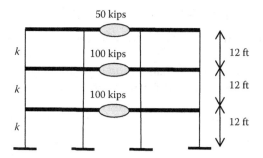

FIGURE E6.1   Schematic of three-story building frame.

$$f_{3o} = \tilde{\Gamma} m_3 \psi_3 A = 1.421(50 \text{ kips}/g)(1)(0.25 \cdot 2.71 \cdot g) = 48.1 \text{ kips}$$

$$f_{2o} = \tilde{\Gamma} m_2 \psi_2 A = 1.421(100 \text{ kips}/g)(2/3)(0.25 \cdot 2.71 \cdot g) = 64.2 \text{ kips}$$

$$f_{1o} = \tilde{\Gamma} m_1 \psi_1 A = 1.421(100 \text{ kips}/g)(1/3)(0.25 \cdot 2.71 \cdot g) = 32.1 \text{ kips}$$

6. Determine base shear and overturning moment: Determine the base shear force using Equation 6.40,

$$V_{\max b} = \sum_{j=1}^{3} f_{jo} = 32.1 \text{ kips} + 64.2 \text{ kips} + 48.1 \text{ kips} = 144.4 \text{ kips}$$

Determine the overturning moment using Equation 6.41,

$$M_{\max b} = \sum_{j=1}^{3} h_j f_{oj} = 32.1 \text{ kips}(12 \text{ ft}) + 64.2 \text{ kips}(24 \text{ ft}) + 48.1 \text{ kips}(36 \text{ ft}) = 3658 \text{ kip ft}$$

We can also obtain the internal story shear forces and moments at each level using Equations 6.38 and 6.39, respectively. Alternatively, we can draw shear force and bending moment diagrams by treating the building as a cantilever beam in order to obtain the internal story shear forces and moments shown in Figure E6.2. This diagram also summarizes the results of the analysis.

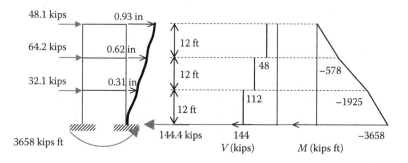

FIGURE E6.2   Lateral forces, internal shear, and internal moment diagram.

6.2   GENERALIZED SDOF CONTINUOUS SYSTEM

The equation of motion for a cantilever tower (or beam) with distributed mass, $m(x)$, and distributed flexural stiffness, $EI(x)$, can be formulated using the tower shown in Figure 6.4. The deformed shape of the system along with the virtual deformation used in the formulation of the equation of motion is depicted in the figure. In this case, the deformation of the tower along its length is $u(x,t)$, which can be decoupled into two functions, one in terms of position, $\psi(x)$, and the other in terms of time, $z(t)$, representing the dynamic displacement at the top of the tower. Also shown in the figure is an equivalent generalized SDOF.

6.2.1   EQUATION OF MOTION

The equation of motion for a tower subjected to a time-dependent input force or ground excitation can be formulated in terms of $z(t)$ only, provided we can find an appropriate shape of the deformed system as it vibrates, $\psi(x)$. The generalized equation of motion in terms of $z(t)$ can again be derived using the principle of virtual work as described in the last section; the resulting second-order ordinary differential equation is of the same form as Equation 6.29 and is rewritten below for convenience,

$$\tilde{m}\ddot{z} + \tilde{c}\dot{z} + \tilde{k}z = -\tilde{L}\ddot{u}_g(t)$$

In the continuous system, integration is used to determine properties of the generalized SDOF as follows:

$$\tilde{m} = \int_0^L m(x)[\psi(x)]^2\, dx \text{ is the generalized mass}$$

$$\tilde{k} = \int_0^L EI(x)[\psi''(x)]^2\, dx \text{ is the generalized stiffness}$$

$$\tilde{L} = \int_0^L m(x)\psi(x)\, dx \text{ is the generalized force}$$

Generalized damping, $\tilde{c}$, is not explicitly determined; it's estimated using the damping ratio, L is the tower height,

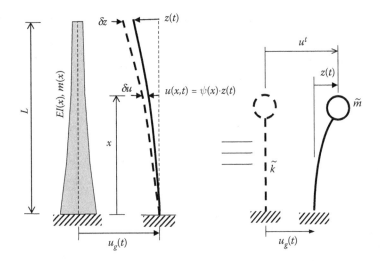

FIGURE 6.4   Tower with distributed properties and equivalent generalized SDOF system.

$m(x)$ is the mass per unit height at an arbitrary height x,
$EI(x)$ is the flexural stiffness per unit height at an arbitrary height x, and
$\psi''(x) = d^2\psi/dx^2$, the second derivative of $\psi(x)$ with respect to x.

Again, this generalized equation of motion is the same as the one that characterizes the vibration of the SDOF oscillator introduced in Chapter 3; thus, all analysis procedures described so far can be used to determine the response $z(t)$. With $z(t)$ and the shape function $\psi(x)$, we can determine a function of the shape into which the tower vibrates, which is the relative displacement at an arbitrary height x,

$$u(x,t) = \psi(x)z(t) \tag{6.42}$$

The accuracy of the results in this case depends on how close the assumed shape function tracks the actual shape of the deformed tower. A satisfactory approximation for the shape function, $\psi(x)$, must satisfy the geometric boundary conditions of the tower, which for this case include zero ground lateral relative displacement and rotation; that is, $u(0) = 0$ and $u'(0) = 0$, where the prime notation indicates differentiation with respect to x, $u'(x) = du/dx$. Also, since z is only a function of t, these boundary conditions reduce to $\psi(0) = 0$ and $\psi'(0) = 0$. Furthermore, the function $\psi(x)$ must be normalized at the top of the tower such that $\psi(L) = 1.0$, where L is the height of the tower.

This analysis can be used to determine internal stress resultants in the tower, which can then be used to determine the stresses. The analysis procedure can be used with a variety of cantilever systems subjected to different time-dependent loads (Chopra 2012). Like the shear building, here we assume a tower subjected to a time-dependent ground lateral displacement of $u_g(t)$ as shown in Figure 6.4, which gives a total lateral displacement at a height x of

$$u^t(x,t) = u(x,t) + u_g(t) \tag{6.43}$$

6.2.2 GENERALIZED PARTICIPATION FACTOR, FREQUENCY, AND NATURAL PERIOD

Rearranging Equation 6.43 into the form of Equation 3.1, we obtain the same results as Equation 6.30 in Section 6.1.2, except that the generalized properties are now computed using the integral relationships. That is,

$$\ddot{z} + 2\zeta\omega_n\dot{z} + \omega_n^2 z = -\tilde{\Gamma}\ddot{u}_g(t)$$

where:

$\omega_n = \sqrt{\dfrac{\tilde{k}}{\tilde{m}}}$ is the generalized natural frequency

$\tilde{\Gamma} = \dfrac{\tilde{L}}{\tilde{m}}$ is the generalized participation factor

ζ is an estimate of the damping ratio

generalized natural period is given as before, $T_n = 2\pi\sqrt{\dfrac{\tilde{m}}{\tilde{k}}}$

6.2.3 DEFLECTIONS, BASE SHEAR, AND MOMENTS USING
PARTICIPATION FACTORS AND RESPONSE SPECTRA

As discussed in Section 6.1.3, the maximum response of a generalized SDOF system due to a dynamic excitation (or force) can be determined using the generalized period and an appropriate design response spectrum. The maximum displacement at the top of the tower based on response spectrum analysis is

$$z = \tilde{\Gamma} \cdot D \tag{6.44}$$

where:

 D is the spectral displacement

The maximum dynamic displacement relative to the ground along the entire tower can be obtained using this displacement

$$u_o(x) = z \cdot \psi(x) = \tilde{\Gamma} \cdot D \cdot \psi(x) \tag{6.45}$$

This can also be written in terms of the spectral acceleration, A, as

$$u_o(x) = \tilde{\Gamma} \cdot \frac{A}{\omega_n^2} \cdot \psi(x) \tag{6.46}$$

The acceleration is approximately proportional to the displacement; that is, $A \cong \omega_n^2 D$.

The equivalent static forces associated with these floor displacements can be obtained using the product of the maximum displacement and associated flexural stiffness, or the product of the mass and maximum total acceleration; that is,

$$f_o(x,t) = m(x) \cdot \psi(x) \cdot \tilde{\Gamma} \cdot A \tag{6.47}$$

This can also be written in terms of the spectral displacement, D, as

$$f_o(x,t) = \omega_n^2 \cdot m(x) \cdot \psi(x) \cdot \tilde{\Gamma} \cdot D \tag{6.48}$$

The internal shear force, V_{max}, and internal story moment, M_{max}, at an arbitrary height, x, can be obtained by applying static equilibrium,

$$V_{max}(x) = \int_x^L f_o(\xi)d\xi = \tilde{\Gamma}A\int_x^L m(\xi)\psi(\xi)d\xi \tag{6.49}$$

$$M_{max}(x) = \int_x^L (\xi - x)f_o(\xi)d\xi = \tilde{\Gamma}A\int_x^L (\xi - x)m(\xi)\psi(\xi)d\xi \tag{6.50}$$

where:

 ξ ranges from x to L

Setting x equal to 0 in Equations 6.49 and 6.50 gives the base shear force, $V_{max\,b}$, and overturning moment, $M_{max\,b}$,

$$V_{max\,b} = \tilde{\Gamma}A\tilde{L} \tag{6.51}$$

$$M_{\max b} = \tilde{\Gamma} A \int_0^L xm(x)\psi(x)dx \qquad (6.52)$$

EXAMPLE 6.2

The reinforced concrete chimney shown in Figure E6.3 is subjected to a ground acceleration due to an earthquake characterized by an 84.1% design spectrum, scaled to 0.25g PGA. Determine (i) period, (ii) peak displacement, (iii) maximum base shear, and (iv) maximum overturning moment of the chimney. Assume diameter, $d = 3.0$ ft; thickness $= 4$ in; modulus of elasticity, $E_c = 3600$ ksi; concrete weight, $\gamma_c = 150$ pcf; damping, $\zeta = 5\%$; and the shape function shown in Figure E6.3.

Solution

1. Determine the properties of the chimney: The properties per unit length at an arbitrary height, x, are as follows:
   Cross-sectional area at x,

   $$\text{Area}(x) = \pi(R_o)^2 - \pi(R_i)^2 = \pi[(1.5 \text{ ft})^2 - (1.167 \text{ ft})^2] = 2.79 \text{ ft}^2$$

   Mass at x per unit foot of tower,

   $$m(x) = \text{Area}(x) \cdot \gamma_c / g = 2.79 \text{ ft}^2 (150 \text{ pcf}) / 32.2 \text{ ft/s}^2 = 13.0 (\text{lbs} \cdot \text{s}^2/\text{ft})/\text{ft}$$

   Moment of inertia at x,

   $$I(x) = \pi/4 \cdot (R_o)^4 - \pi/4 \cdot (R_i)^4 = \pi/4[(1.5 \text{ ft})^4 - (1.167 \text{ ft})^4] = 2.52 \text{ ft}^4$$

   Flexural stiffness at x

   $$E_c I(x) = 3600 \text{ ksi} (1000 \text{ lb/kip})(12 \text{ in/ft})^2 (2.52 \text{ ft}^4) = 1.307 \times 10^9 \text{ lbs} \cdot \text{ft}^2$$

2. Determine the generalized properties: The generalized mass, stiffness, and force of the generalized SDOF system are calculated as follows:
   Generalized mass,

   $$\tilde{m} = \int_0^L m(x)[\psi(x)]^2 dx = 13 \text{ lb s}^2/\text{ft}^2 \int_0^{30 \text{ ft}} \left[\frac{x^2}{(30 \text{ ft})^2} \right]^2 dx = 78.1 \text{ lb s}^2/\text{ft}$$

   Generalized stiffness,

   $$\psi''(x) = \frac{d^2\psi(x)}{dx^2} = \frac{2}{(30 \text{ ft})^2}$$

   $$\tilde{k} = \int_0^L EI(x)[\psi''(x)]^2 dx = 1.307 \times 10^9 \text{ lbs} \cdot \text{ft}^2 \int_0^{30 \text{ ft}} \left[\frac{2}{(30 \text{ ft})^2} \right]^2 dx = 1.94 \times 10^5 \text{ lb/ft}$$

The generalized force is obtained using

$$\tilde{L} = \int_0^L m(x)\psi(x)dx = \frac{13\,\text{lb s}^2/\text{ft}^2}{(30\,\text{ft})^2}\int_0^{30\,\text{ft}} x^2 dx = 130\,\text{lb s}^2/\text{ft}$$

3. Determine the natural period, frequency, and participation factor of the generalized SDOF system.

   Natural period,

$$T_n = 2\pi\sqrt{\frac{\tilde{m}}{\tilde{k}}} = 2\pi\sqrt{\frac{78.1\,\text{lb}-\text{s}^2/\text{ft}}{1.94\times10^5\,\text{lb/ft}}} = 0.126\,\text{s}$$

   Natural frequency,

$$\omega_n = \sqrt{\frac{\tilde{k}}{\tilde{m}}} = \sqrt{\frac{1.94\times10^5\,\text{lb/ft}}{78.1\,\text{lb}-\text{s}^2/\text{ft}}} = 49.8\,\text{rad/s}$$

   Participation factor,

$$\tilde{\Gamma} = \frac{\tilde{L}}{\tilde{m}} = \frac{130\,\text{lb}-\text{s}^2/\text{ft}}{78.1\,\text{lb}-\text{s}^2/\text{ft}} = 1.667$$

4. Determine maximum displacement at the top of the tower: This displacement is $z(t)$ or D due to 84.1% design spectrum, scaled to 0.25g PGA; first, enter the response spectrum in Figure 5.11 with $T_n = 0.126$ s, which is greater than 1/8 s, so the acceleration is constant,

$$A = \eta \cdot 2.71g = 0.25(2.71)(386.4\,\text{in/s}^2) = 261.8\,\text{in/s}^2 = 21.8\,\text{ft/s}^2$$

   where:
     η is the PGA scale for a given earthquake; 0.25 in this case

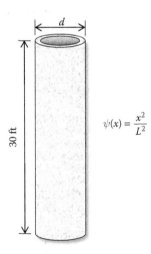

FIGURE E6.3   Schematic of a reinforced concrete chimney.

The maximum spectral displacement is

$$D = \frac{A}{\omega_n^2} = \frac{261.8 \text{ in/s}^2}{(49.8 \text{ rad/s})^2} = 0.106 \text{ in}$$

We can now determine the maximum displacement at the top of the tower,

$$u_o = \tilde{\Gamma} \cdot D = (1.667)(0.106 \text{ in}) = 0.176 \text{ in}$$

5. Determine base shear and overturning moment: Determine the base shear force using Equation 6.51,

$$V_{\text{max } b} = \tilde{\Gamma} A \tilde{L} = 1.667(261.8 \text{ in/s}^2)(130 \text{ lb s}^2/\text{ft})(\text{ft}/12 \text{ in})(\text{kip}/1000 \text{ lb}) = 4.73 \text{ kips}$$

Determine the overturning moment using Equation 6.52

$$M_{\text{max } b} = \tilde{\Gamma} A \int_0^L xm(x)\psi(x)dx = 1.667(21.8 \text{ ft/s}^2)\frac{13 \text{ lb s}^2/\text{ft}^2}{(30 \text{ ft})^2} \int_0^{30\text{ft}} x^3 dx = 106 \text{ kip} \cdot \text{ft}$$

PROBLEMS

6.1 Solve Example 6.1 with a shape function given for a low-rise building.

6.2 Solve Example 6.1 with a shape function of $\psi(x) = (x/H)^2$, where H is the building height.

6.3 Use the generalized SDOF analysis for the following building frame to determine the story displacements, story forces, base shear, and overturning moment due to an earthquake characterized by an 84.1% design spectrum, scaled to 0.25g PGA. Let us assume damping of 5% and a shape function of $\psi(x) = \sin(\pi x/2H)$, where H is the building height.

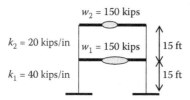

6.4 Use the generalized SDOF analysis for the following building frame to determine the story displacements, story forces, base shear, and overturning moment due to an earthquake characterized by an 84.1% design spectrum, scaled to 0.25g PGA. Let us assume damping of 5% and a shape function given for a low-rise building.

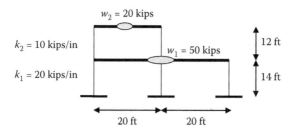

6.5   Given a five-story building frame subjected to a ground acceleration due to a 0.25g PGA earthquake characterized by the 84.1% design spectrum, use generalized SDOF analysis to determine (i) peak displacements, (ii) maximum base shear, and (iii) maximum floor overturning moments. $k = 326.3$ kips/in, $w_5 = 50$ kips, and the weights on the other four floors equal 100 kips. Let us assume rigid beams, damping of 5%, and a shape function given for a mid-rise building.

6.6   Solve Problem 6.5 with a shape function given for a high-rise building.

6.7   Solve Example 6.2 with a shape function of $\psi(x) = 1.5\,(x/L)^2 - 0.5\,(x/L)^3$.

6.8   Solve Example 6.2 with a shape function of $\psi(x) = 1 - \cos(\pi x/2L)$.

REFERENCES

Chopra, A. K., *Dynamics of Structures: Theory and Applications to Earthquake Engineering*, 4th edition, Prentice-Hall, Upper Saddle River, NJ, 2012.

Naeim, F., *The Seismic Design Handbook*, 2nd edition, Springer, Dordrecht, The Netherlands, 2001.

7 Multi-Degree-of-Freedom System Analysis

After reading this chapter, you will be able to:

1. Setup the equation of motion for an multi-degree of freedom (MDOF) system
2. Determine frequencies and periods for an MDOF system
3. Determine participation factors for an MDOF system
4. Use response spectra and participation factors to calculate the maximum shear and moment response of an MDOF system

In Chapter 6, we modeled multi-degree-of-freedom (MDOF) systems as generalized single-degree-of-freedom (SDOF) systems to obtain the dynamic response caused by a ground excitation. This analysis procedure is convenient for structures that can assume a unique shape during vibration. However, practical structures can assume several different shapes (mode shapes) during their vibration (see Figure 7.1), with each mode contributing to the total response of the structure. Also, each mode shape shown in Figure 7.1 has a unique associated frequency, with the first (or fundamental) mode of vibration defined as having the lowest frequency, or the longest period. The other modes have shorter periods and are known as higher modes, or harmonics. Depending on the frequency content of the time-dependent excitation, different mode shapes may contribute differently to the total response of the system; typically, the first or fundamental mode has the greatest influence on the response, but because of resonance other modes can dominate the response at higher harmonics.

Since each individual mode of vibration has its own period, they each can be represented by a generalized SDOF system of the same period (following the same process as in Chapter 6). Also, each mode shape remains constant regardless of the amplitude of the displacement. Thus, each mode shape can be assigned a reference unit amplitude value at a specified location, say the top of the structure as in Chapter 6. The actual amplitude can then be obtained from the dynamic analysis of the generalized SDOF. Therefore, once we obtain equations or vectors representing the mode shapes and associated periods, the maximum response for the separate modes can be obtained by analyzing each mode as a distinct generalized SDOF system. However, it is not likely that the maximum response values will all occur simultaneously at the same level; therefore, the maximum responses from all generalized SDOF cases are combined statistically in order to obtain the total system response. A number of procedures have been developed to combine the maximum results; two methods will be discussed later in the chapter, square root of the sum of the squares (SRSS) or the complete quadratic combination (CQC) methods, both of which are mentioned in the code-based dynamic seismic analysis procedures.

All modes for shear buildings such as the one shown in Figure 7.2 can be modeled simultaneously using matrix methods; this is known as modal analysis and will be the subject of the remainder of the chapter. For example, the frame shown in Figure 7.2 has four degrees of freedom and thus four mode shapes. Each mode shape has a differential equation associated with it. For this discrete case, a system of four dependent differential equations is formulated. The process entails decoupling the system of differential equations into four independent equations, which can then be used to find maximum response values using a response spectrum. The key concept to obtaining the mode shapes and the associated periods is a solution to an eigenvalue problem; the eigenvalues are related to the periods and the corresponding eigenvectors give the mode shapes.

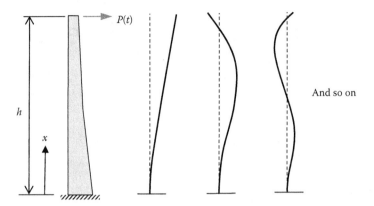

FIGURE 7.1 Structural mode shapes.

The eigen matrix operation can be viewed as a linear transformation that maps vectors into multiples of themselves. That is,

$$\mathbf{A}\mathbf{x} = \lambda\mathbf{x} \quad \text{or} \quad [A]\{x\} = \lambda\{x\} \tag{7.1}$$

where λ is the scalar proportionality constant.

A must be a square matrix, with as many values of λ as there are elements in each row. This equation can also be rewritten as

$$(\mathbf{A} - \lambda\mathbf{I})\mathbf{x} = \mathbf{0} \quad \text{or} \quad ([A] - \lambda[I])\{x\} = \{0\} \tag{7.2}$$

where:

I is the identify matrix

The nonzero solutions to this equation require that λ be chosen so that

$$\det(\mathbf{A} - \lambda\mathbf{I}) = 0 \quad \text{or} \quad \det([A] - \lambda[I]) = 0 \tag{7.3}$$

The values of λ that satisfy this relationship correspond to eigenvalues of matrix **A**; each λ yields an eigenvector $\{x\}$ solution of Equation 7.1 to within a multiplicative constant. That is, scaling the eigenvector by any constant factor will still satisfy Equation 7.1. Consequently, each eigenvector is normalized with respect to one of its elements, usually the first element.

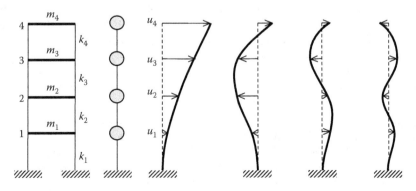

FIGURE 7.2 Idealized four-story shear building system and mode shapes.

These solutions are practically intractable via hand calculations for systems with more than two degrees of freedom, which is why solutions are regularly found via computer using one of the widely available numerical packages, such as MATLAB®. With this program, we can use the eig operator to determine the eigenvectors denoted as *phi* and the eigenvalues denoted as *lambda* as follows,

$$[\texttt{phi, lambda}] = \texttt{eig(A)} \tag{7.4}$$

In the sections that follow, we present the application of the eigenvalue problem to the solution of structural dynamics problems. This solution will then be extended to encompass the calculation of the maximum shear and moment response of MDOF systems.

7.1 EQUATIONS OF MOTION FOR AN MDOF SYSTEM

To formulate the equation of motion for an MDOF system, we first use the free vibration response of the two-degree-of-freedom shear building shown in Figure 7.3, and then expand it to an *N*-degree-of-freedom system. The solution to the resulting equation of motion yields the natural periods (or frequencies) and the corresponding vibration mode shapes. Recall from Section 3.1.1 that free vibration is caused by initial conditions (displacement and velocity), not a time-dependent applied force. To characterize the free vibration motion of an MDOF system, we need to establish a system of equations of motion for the model in terms of relative displacements (for stiffness forces) and total displacements (for inertial forces). These equations can be determined by applying equilibrium to the free-body diagrams of the masses in Figure 7.3 using D'Alembert's principle. Notice that we have not included vertical forces or internal moments in the column stems because we only need to apply horizontal equilibrium to obtain the equations of motion.

Applying horizontal equilibrium to the free-body diagrams of the two masses shown in Figure 7.3 yields two equations of motion.

Equilibrium of mass 1:

$$+\rightarrow \sum F_x = 0; \quad -m_1\ddot{u}_1 - k_1u_1 + k_2(u_2 - u_1) = 0 \Rightarrow m_1\ddot{u}_1 + (k_1 + k_2)u_1 - k_2u_2 = 0 \tag{7.5}$$

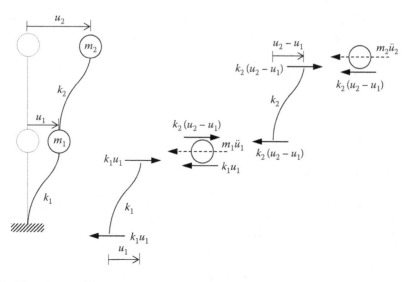

FIGURE 7.3 Two-degree-of-freedom system and FBD of the two masses.

where m_1 is the mass of the first floor, which can be determined following the procedure discussed in Section 3.1.2, and k_1 and k_2 are the story stiffnesses of the first and second levels, respectively, which can be determined by adding the lateral stiffnesses of all columns in the story. Each story can be treated as a portal frame as discussed in Chapter 3; thus, for a story of height h, column flexural stiffness EI, and assuming rigid floor diaphragms (rigid frame beams), the lateral stiffness of a column is $12EI/h^3$ and the stiffness of story j is $k_j = \sum_{\text{columns}} 12EI/h^3$.

Equilibrium of mass 2:

$$+ \rightarrow \sum F_x = 0; \quad -m_2\ddot{u}_2 - k_2(u_2 - u_1) = 0 \Rightarrow m_2\ddot{u}_2 - k_2u_1 + k_2u_2 = 0 \tag{7.6}$$

where m_2 is the mass of the second floor.

These two dependent equations of motion (Equations 7.5 and 7.6) are coupled second-order, linear, and homogeneous differential equations with constant coefficients, and can be rewritten in matrix form as

$$\begin{bmatrix} m_1 & 0 \\ 0 & m_2 \end{bmatrix} \begin{Bmatrix} \ddot{u}_1 \\ \ddot{u}_2 \end{Bmatrix} + \begin{bmatrix} k_1 + k_2 & -k_2 \\ -k_2 & k_2 \end{bmatrix} \begin{Bmatrix} u_1 \\ u_2 \end{Bmatrix} = \begin{Bmatrix} 0 \\ 0 \end{Bmatrix} \tag{7.7}$$

In general,

$$[m]\{\ddot{u}\} + [k]\{u\} = \{0\} \tag{7.8}$$

where $[m]$ is the mass matrix, which is a diagonal matrix (nonzero elements only along the forward diagonal), and $[k]$ is the stiffness matrix; this can alternatively be obtained by directly finding each coefficient k_{ij}, which represents a force needed at level i to hold a unit displacement at level j, holding zero displacements at all other levels.

Notice that an Nth-degree-of-freedom system results in N-dependent equations, or $N \times N$ mass and stiffness matrices.

7.1.1 PERIODS AND MODE SHAPES FOR AN MDOF SYSTEM

As discussed in the last section, and SDOF analysis in Chapter 3, the free-vibration response yields the natural periods (or frequencies). Since this is similar to the SDOF case, we can assume that the resulting motion is described by a simple harmonic equation with the following form for displacement:

$$\{u\} = \{a\}\sin(\omega_n t - \alpha) \tag{7.9}$$

where:

$\omega_n = \sqrt{k/m}$ is the natural circular frequency with units of radians per second (rad/s), which can be related to the period as $T_n = 2\pi/\omega_n$

Taking the first and second derivatives of Equation 7.9 with respect to time,

$$\{\dot{u}\} = \{a\}\omega_n \cos(\omega_n t - \alpha) \tag{7.10}$$

$$\{\ddot{u}\} = -\{a\}\omega_n^2 \sin(\omega_n t - \alpha) = -\omega_n^2\{u\} \tag{7.11}$$

Substituting these equations into Equation 7.8 yields

$$[m]\left(-\omega_n^2\{u\}\right)+[k]\{u\}=\{0\} \tag{7.12}$$

Factoring out the displacement vector $\{u\}$, we get

$$\left[[k]-\omega_n^2[m]\right]\{u\}=\{0\} \tag{7.13}$$

This equation is similar to Equation 7.2 and has a nontrivial solution (one that gives nonzero values for $\{u\}$) if Equation 7.3 is satisfied. That is,

$$\det\left([k]-\omega_n^2[m]\right)=0 \tag{7.14}$$

This is the characteristic (or eigen) equation of the system, also known as the frequency equation because expanding the determinate results in an Nth degree polynomial in terms of ω_n^2. That is, Equation 7.14, in general, should be satisfied for N values of ω_n^2. We can then solve the homogeneous system of equations (Equation 7.12) for $a_1, a_2, \ldots, a_N$ for each value of ω_n^2, totaling N modes of vibration.

Solving this eigenvalue problem is convenient using a computer program such as MATLAB; for this case, the eig operator includes the stiffness, $[k]$ and the mass, $[m]$ matrices to determine eigenvectors, denoted by *phi* and eigenvalues denoted by *lambda*, as

$$[phi, lambda] = eig(k,m) \tag{7.15}$$

For illustration purposes, let us conduct the analysis of the two-degree-of-freedom case in Figure 7.3 using hand calculations. Recall that the assumed displacement response vector is given as

$$\begin{Bmatrix} u_1 \\ u_2 \end{Bmatrix} = \begin{Bmatrix} a_1 \\ a_2 \end{Bmatrix} \sin(\omega_n t - \alpha) \tag{7.16}$$

Taking two derivatives of Equation 7.16 with respect to time and substituting into Equation 7.7 gives the following equation:

$$-\omega_n^2 \begin{bmatrix} m_1 & 0 \\ 0 & m_2 \end{bmatrix} \begin{Bmatrix} a_1 \\ a_2 \end{Bmatrix} + \begin{bmatrix} k_1+k_2 & -k_2 \\ -k_2 & k_2 \end{bmatrix} \begin{Bmatrix} a_1 \\ a_2 \end{Bmatrix} = \begin{Bmatrix} 0 \\ 0 \end{Bmatrix} \tag{7.17}$$

which can be rewritten as

$$\begin{bmatrix} k_1+k_2-m_1\omega_n^2 & -k_2 \\ -k_2 & k_2-m_2\omega_n^2 \end{bmatrix} \begin{Bmatrix} a_1 \\ a_2 \end{Bmatrix} = \begin{Bmatrix} 0 \\ 0 \end{Bmatrix} \tag{7.18}$$

As discussed above, the trivial solution $a_1 = a_2 = 0$ does not provide any information, and the nontrivial solution requires that the determinate of the matrix be zero; that is,

$$\det \begin{bmatrix} k_1+k_2-m_1\omega_n^2 & -k_2 \\ -k_2 & k_2-m_2\omega_n^2 \end{bmatrix} = 0 \tag{7.19}$$

Expanding this determinate results in a second-order polynomial in ω_n^2, the frequency equation,

$$m_1 m_2 \omega_n^4 - ((k_1 + k_2)m_2 + m_1 k_2)\omega_n^2 + k_1 k_2 = 0 \tag{7.20}$$

We simplify the analysis further by assuming $k_1 = k_2 = k$ and $m_1 = m_2 = m$, which leads to the following polynomial:

$$m^2 \omega_n^4 - 3km\omega_n^2 + k^2 = 0 \tag{7.21}$$

We use the quadratic formula, $x = (-b \pm \sqrt{b^2 - 4ac})/2a$, to find the roots of Equation 7.21, which are the eigen (characteristic) values:

$$\omega_n^2 = \frac{3km \pm \sqrt{(3km)^2 - 4m^2 k^2}}{2m^2} = \frac{3 \pm \sqrt{5}}{2}\frac{k}{m} \tag{7.22}$$

or

$$\omega_{n1}^2 = \frac{3 - \sqrt{5}}{2}\frac{k}{m} \quad \text{and} \quad \omega_{n2}^2 = \frac{3 + \sqrt{5}}{2}\frac{k}{m} \tag{7.23}$$

The corresponding frequencies are

$$\omega_{n1} = 0.618\sqrt{\frac{k}{m}} \quad \text{and} \quad \omega_{n2} = 1.618\sqrt{\frac{k}{m}} \tag{7.24}$$

Frequencies are usually listed in ascending order; that is, from the smallest to largest.

Also, each of these eigenvalues is associated with a distinct eigenvector (or mode shape). The actual values of the eigenvector elements in Equation 7.18, a_1 and a_2 for this case, are indeterminate. However, each eigenvector can be obtained by setting the values of the top element equal to one, and solving for the values of the other eigenvector elements relative to top value. Substituting ω_{n1}^2 into Equation 7.18 yields

$$\begin{bmatrix} 2k - m\left(\dfrac{3 - \sqrt{5}}{2}\dfrac{k}{m}\right) & -k \\[2ex] -k & k - m\left(\dfrac{3 - \sqrt{5}}{2}\dfrac{k}{m}\right) \end{bmatrix} \begin{Bmatrix} a_1 \\ a_2 \end{Bmatrix} = \begin{Bmatrix} 0 \\ 0 \end{Bmatrix} \tag{7.25}$$

Simplifying the elements of this matrix and setting $a_2 = 1$, we get

$$k\begin{bmatrix} 1.618 & -1 \\ -1 & 0.618 \end{bmatrix}\begin{Bmatrix} a_1 \\ a_2 \end{Bmatrix} = \begin{Bmatrix} 0 \\ 0 \end{Bmatrix} \quad \text{or} \quad \begin{bmatrix} 1.618 & -1 \\ -1 & 0.618 \end{bmatrix}\begin{Bmatrix} a_1 \\ 1 \end{Bmatrix} = \begin{Bmatrix} 0 \\ 0 \end{Bmatrix} \tag{7.26}$$

Solving for a_1, the resulting eigenvector (mode shape) is

$$\begin{Bmatrix} \phi_{11} \\ \phi_{21} \end{Bmatrix} = \begin{Bmatrix} a_1 \\ a_2 \end{Bmatrix}_1 = \begin{Bmatrix} 0.618 \\ 1 \end{Bmatrix} \tag{7.27}$$

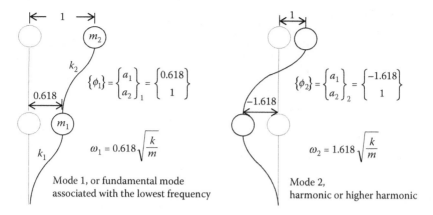

FIGURE 7.4   Mode shapes and associated frequencies.

Similarly, we can obtain the second eigenvector (mode shape) as

$$\begin{Bmatrix} \phi_{12} \\ \phi_{22} \end{Bmatrix} = \begin{Bmatrix} a_1 \\ a_2 \end{Bmatrix}_2 = \begin{Bmatrix} -1.618 \\ 1 \end{Bmatrix} \tag{7.28}$$

These eigenvectors are referred to as *normal modes* because they are normalized so that the displacements of the mass at the top of the structure are equal to one. The resulting eigenvectors describe the mode shapes and are defined as $\{\phi\}_j$ (this was the process followed in Chapter 6 to obtain a shape vector that describes the deflected shape of a vibrating shear building). A summary of these results, including the normal modes and frequencies, is shown in Figure 7.4.

Regardless of how the eigenvectors are normalized, the resulting mode shapes are orthogonal; that is, for two distinct mode shapes of the structure, $r \neq s$,

$$\{\phi_r\}^T \{\phi_s\} = 0 \tag{7.29}$$

which leads to the following relationships:

$$\{\phi_r\}^T [k] \{\phi_s\} = 0 \quad \text{and} \quad \{\phi_r\}^T [m] \{\phi_s\} = 0 \tag{7.30}$$

where the superscript T denotes the transpose matrix operation.

We can now obtain the total solution of the system by combining the two modes as

$$\begin{aligned} u_1 &= \phi_{11}q_1 + \phi_{12}q_2 \\ u_2 &= \phi_{21}q_1 + \phi_{22}q_2 \end{aligned} \tag{7.31}$$

where we now assume a new set of harmonic displacement functions with constants A_j and B_j,

$$\begin{aligned} q_1 &= A_1 \sin \omega_1 t + B_1 \cos \omega_1 t \\ q_2 &= A_2 \sin \omega_2 t + B_2 \cos \omega_2 t \end{aligned} \tag{7.32}$$

Rewriting Equations 7.31 in the matrix form

$$\begin{Bmatrix} u_1 \\ u_2 \end{Bmatrix} - \begin{bmatrix} \phi_{11} & \phi_{12} \\ \phi_{21} & \phi_{22} \end{bmatrix} \begin{Bmatrix} q_1 \\ q_2 \end{Bmatrix} \quad \text{or} \quad \{u\} = [\phi]\{q\} \tag{7.33}$$

The matrix $[\phi]$ is a coordinate transformation matrix (also known as the modal matrix) that maps the original coordinates $\{u\}$ to the *principal coordinates* $\{q\}$. The inverse transformation is

$$\{q\} = [\phi]^{-1}\{u\} \tag{7.34}$$

The orthogonality condition given by Equation 7.29 implies that $[\phi]^{-1} = [\phi]^T$. Thus, we can then rewrite Equation 7.34 as

$$\{q\} = [\phi]^T\{u\} \tag{7.35}$$

Substituting the transformation relationship, Equation 7.33, into the equations of motion, Equation 7.8, and premultiplying by $[\phi]^T$ yields

$$[\phi]^T[m][\phi]\{\ddot{q}\} + [\phi]^T[k][\phi]\{q\} = \{0\} \quad \text{or} \quad [M]\{\ddot{q}\} + [K]\{q\} = \{0\} \tag{7.36}$$

The matrices $[M]$ and $[K]$ are diagonal because of the orthogonality condition given by Equation 7.30 (i.e., all off diagonal elements are zero). These are known as *modal mass* and *modal stiffness* matrices, respectively. The diagonal elements of these matrices are given by

$$M_i = \{\phi_i\}^T[m]\{\phi_i\} = \sum_{j=1}^N m_j \phi_{ji}^2 \quad \text{and} \quad K_i = \{\phi_i\}^T[k]\{\phi_i\} = \sum_{j=1}^N k_j \phi_{ji}^2 \tag{7.37}$$

Notice that Equations 7.36 are now uncoupled, or independent. That is,

$$M_i \ddot{q}_i + K_i q_i = 0 \tag{7.38}$$

If each equation is divided by the associated modal mass, the equations can be rewritten as

$$\ddot{q}_i + \omega_{ni}^2 q_i = 0 \tag{7.39}$$

So, each equation has the same form as the equation of motion for the SDOF system we described in detail in Chapter 3. This implies that we can use the solutions to the different loadings or forcing functions examined previously, including response spectrum analysis.

To illustrate the process of decoupling the equations of motion, let us revisit the two-degrees-of-freedom case introduced earlier in this chapter. Using Equation 7.36, we get the mass transformation coefficients M_1 and M_2:

$$\begin{bmatrix} M_1 & 0 \\ 0 & M_2 \end{bmatrix} = \begin{bmatrix} 0.618 & 1 \\ -1.618 & 1 \end{bmatrix} \begin{bmatrix} m & 0 \\ 0 & m \end{bmatrix} \begin{bmatrix} 0.618 & -1.618 \\ 1 & 1 \end{bmatrix} = \begin{bmatrix} 1.382m & 0 \\ 0 & 3.618m \end{bmatrix} \tag{7.40}$$

and the stiffness coefficients, K_1 and K_2,

$$\begin{bmatrix} K_1 & 0 \\ 0 & K_2 \end{bmatrix} = \begin{bmatrix} 0.618 & 1 \\ -1.618 & 1 \end{bmatrix} \begin{bmatrix} 2k & -k \\ -k & k \end{bmatrix} \begin{bmatrix} 0.618 & -1.618 \\ 1 & 1 \end{bmatrix} = \begin{bmatrix} 0.528k & 0 \\ 0 & 9.472k \end{bmatrix} \quad (7.41)$$

This also proves the orthogonality condition ($\{\phi_1\}^T[m]\{\phi_2\} = \{\phi_2\}^T[m]\{\phi_1\} = \{0\}$ and $\{\phi_1\}^T[k]$ $\{\phi_2\} = \{\phi_2\}^T[k]\{\phi_1\} = \{0\}$) since the off-diagonal elements of the two matrices are zero. That is,

$$\{\phi_1\}^T[m]\{\phi_2\} = [0.618 \quad 1] \begin{bmatrix} m & 0 \\ 0 & m \end{bmatrix} \begin{bmatrix} -1.618 \\ 1 \end{bmatrix} = [0.618m \quad m] \begin{bmatrix} -1.618 \\ 1 \end{bmatrix} = 0 \quad (7.42)$$

The equations of motion are

$$\begin{aligned} 1.382 \cdot m \cdot \ddot{q}_1 + 0.528 \cdot k \cdot q_1 &= 0 \\ 3.618 \cdot m \cdot \ddot{q}_2 + 9.472 \cdot k \cdot q_2 &= 0 \end{aligned} \quad (7.43)$$

EXAMPLE 7.1

For the two-story building frame shown in Figure E7.1, determine (1) mass and stiffness matrices, (2) periods and modal matrix (normalized mode shapes), and (3) the stiffness and modal matrices. Assume beams are rigid. $k = 20$ kips/in and $m = 0.4$ kips·s²/in.

SOLUTION

1. Determine the mass and stiffness matrices. These can be obtained following the process shown in Figure 7.3 by setting $m_1 = m_2 = m$, $k_1 = 2k$, and $k_2 = k$, and applying equilibrium to the free-body diagrams of each mass, resulting in equations similar to Equations 7.5 and 7.6. The two equations can be combined into a matrix equation as shown in Equation 7.7:

$$\begin{bmatrix} m & 0 \\ 0 & m \end{bmatrix} \begin{Bmatrix} \ddot{u}_1 \\ \ddot{u}_2 \end{Bmatrix} + \begin{bmatrix} 3k & -k \\ -k & k \end{bmatrix} \begin{Bmatrix} u_1 \\ u_2 \end{Bmatrix} = \begin{Bmatrix} 0 \\ 0 \end{Bmatrix}$$

The mass and stiffness matrices are

$$[m] = \begin{bmatrix} m & 0 \\ 0 & m \end{bmatrix} = \begin{bmatrix} 0.4 & 0 \\ 0 & 0.4 \end{bmatrix} \text{kip·s}^2/\text{in}$$

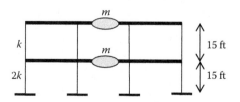

FIGURE E7.1   Schematic of two-story building frame.

$$[k] = \begin{bmatrix} 3k & -k \\ -k & k \end{bmatrix} = \begin{bmatrix} 60 & -20 \\ -20 & 20 \end{bmatrix} \text{kips/in}$$

2. Determine the periods and modal matrix. The natural frequencies and mode shapes can be determined by solving for the eigenvalues and eigenvectors using a standard eigen solution; natural frequencies can be obtained using the eigenvalues and the mode shapes with the eigenvectors. Also, the periods are obtained using the natural frequencies. The frequency equation (Equation 7.20) resulting from the expansion of the eigen equation is

$$m^2 \omega_n^4 - 4 k m \omega_n^2 + 2k^2 = 0$$

Using the quadratic formula, we can find the roots of this equation, which are the eigenvalues:

$$w_n^2 = \frac{4km \pm \sqrt{(4km)^2 - 8m^2k^2}}{2m^2} = \frac{4 \pm 2\sqrt{2}}{2}\frac{k}{m} = \left(2 \pm \sqrt{2}\right)\frac{20 \text{ kips /in}}{0.4 \text{ kip} \cdot s^2/\text{in}}$$

The frequencies are (from smallest to largest)

$$\omega_{n1} = 5.41 \text{ rad/s} \quad \text{and} \quad \omega_{n2} = 13.1 \text{ rad/s}$$

The natural periods, obtained using $T_{nj} = 2\pi/\omega_{nj}$, are

$$T_{n1} = 1.16 \text{ s} \quad \text{and} \quad T_{n2} = 0.48 \text{ s}$$

To obtain mode shapes, we substitute each eigenvalue (ω_{n1}^2 first) into Equation 7.18 and set $a_2 = 1$,

$$\begin{bmatrix} 3k - \left(2-\sqrt{2}\right)k & -k \\ -k & k - \left(2-\sqrt{2}\right)k \end{bmatrix}\begin{Bmatrix} a_1 \\ a_2 \end{Bmatrix} = \begin{Bmatrix} 0 \\ 0 \end{Bmatrix} \quad \text{or} \quad \begin{bmatrix} 2.414 & -1 \\ -1 & 0.414 \end{bmatrix}\begin{Bmatrix} a_1 \\ 1 \end{Bmatrix} = \begin{Bmatrix} 0 \\ 0 \end{Bmatrix}$$

Solving for a_1, the resulting eigenvector (mode shape) is

$$\begin{Bmatrix} \phi_{11} \\ \phi_{21} \end{Bmatrix} = \begin{Bmatrix} a_1 \\ a_2 \end{Bmatrix}_1 = \begin{Bmatrix} 0.414 \\ 1 \end{Bmatrix}$$

Similarly, we can obtain the second eigenvector (mode shape) as

$$\begin{Bmatrix} \phi_{12} \\ \phi_{22} \end{Bmatrix} = \begin{Bmatrix} a_1 \\ a_2 \end{Bmatrix}_2 = \begin{Bmatrix} -2.414 \\ 1 \end{Bmatrix}$$

The modal matrix is

$$[\phi] = \begin{bmatrix} 0.414 & -2.414 \\ 1 & 1 \end{bmatrix}$$

3. Determine the modal mass and stiffness matrices. Use Equations 7.37 to determine the elements of the modal mass and stiffness matrices. Or, we can use the matrix operations in Equation 7.36:
   Modal mass matrix:

$$[M] = [\phi]^T [m] [\phi] = \begin{bmatrix} 0.414 & 1 \\ -2.414 & 1 \end{bmatrix} \begin{bmatrix} 0.4 & 0 \\ 0 & 0.4 \end{bmatrix} \begin{bmatrix} 0.414 & -2.414 \\ 1 & 1 \end{bmatrix}$$

$$= \begin{bmatrix} 0.469 & 0 \\ 0 & 2.73 \end{bmatrix} \text{kip} \cdot \text{s}^2/\text{in}$$

Modal stiffness matrix:

$$[K] = [\phi]^T [k] [\phi] = \begin{bmatrix} 0.414 & 1 \\ -2.414 & 1 \end{bmatrix} \begin{bmatrix} 60 & -20 \\ -20 & 20 \end{bmatrix} \begin{bmatrix} 0.414 & -2.414 \\ 1 & 1 \end{bmatrix}$$

$$= \begin{bmatrix} 13.7 & 0 \\ 0 & 466.3 \end{bmatrix} \text{kips/in}$$

4. Alternatively, we can write a MATLAB script to perform all the operations; the most important of the operations being the eig operator in Equation 7.15:

```
clear all % clears any previously defined variables
clc % clears the screen
m = [0.4 0; 0 0.4]; % mass matrix
k = [60 -20; -20 20]; % stiffness matrix
[phi,lam]=eig(k,m); % compute eigenvalues and eigenvectors
omegas=sqrt(lam) % determine and show frequencies from eigenvalues
periods = 2*pi*diag(inv(omegas)) % determine the periods from omegas
% mode shapes by normalizing eigenvectors to get top displ equal to 1
[N_rows, N_cols] = size(phi); % finds N, the size of the matrices
for i = 1:N_cols; % loops over the N modes to normalize them
    norm_phi(:,i) = phi(:,i)./phi(N_rows,i);
end
norm_phi % display normalized modal matrix
% determine the modal mass and stiffness matrices
M = norm_phi'*m*norm_phi
K = norm_phi'*k*norm_phi
```

The results of this script are the same as those of parts 1–3:

```
omegas =
   5.4120        0
        0  13.0656
periods =
   1.1610
   0.4809
norm_phi =
   0.4142   -2.4142
   1.0000    1.0000
M =
   0.4686    0.0000
   0.0000    2.7314
K =
   13.7258 0.0000
        0 466.2742
```

7.1.2 Participation Factors for an MDOF System

As in Chapter 6, we can account for ground excitation loading in the formulation of the equations of motion for MDOF systems by determining the total displacement response of the system at level j, which is a combination of the relative displacement, $u_j(t)$ and the ground level lateral displacement, $u_g(t)$, caused by seismic ground motion:

$$u_j^t(t) = u_j(t) + u_g(t) \tag{7.44}$$

For MDOF systems, we can substitute the second derivative of Equation 7.44 into the first term of Equation 7.8 to obtain the equations of motion including ground excitation,

$$[m]\{\ddot{u}\} + [k]\{u\} = -[m]\{1\}\ddot{u}_g \tag{7.45}$$

where $\{1\}$ is a column vector of ones.

We can use Equation 7.34, the coordinate transformation matrix $[\phi]$, to transform the original coordinates $\{u\}$ into the *principal coordinates* $\{q\}$. If we substitute the transformation relationship into the equations of motion, and premultiply by $[\phi]^T$, we get

$$[\phi]^T[m][\phi]\{\ddot{q}\} + [\phi]^T[k][\phi]\{q\} = -[\phi]^T[m]\{1\}\ddot{u}_g \tag{7.46}$$

Substituting modal mass and stiffness relations, Equations 7.37, into Equation 7.46, we get

$$[M]\{\ddot{q}\} + [K]\{q\} = -[\phi]^T[m]\{1\}\ddot{u}_g \tag{7.47}$$

Alternatively, these equations of motion can be written as N uncoupled equations as

$$\ddot{q}_j + \omega_{nj}^2 q_j = -\Gamma_j \ddot{u}_g \tag{7.48}$$

which after including the effects of damping can be written as

$$\ddot{q}_j + 2\zeta\omega_{nj}\dot{q}_j + \omega_n^2 q_j = -\Gamma_j \ddot{u}_g \tag{7.49}$$

where the damping is assumed to be constant for all mode shapes. For elastic analysis of MDOF systems, where damping effects are significant, there are a number of approaches used to establish the damping matrix, including a relationship based on the stiffness and mass matrices. This can be found in conventional structural dynamics texts such as Chopra (2012) and Villaverde (2009). Also, we can now define the participation factor for each mode shape (the *modal participation factor*) as

$$\Gamma_j = \frac{L_j}{M_j} = \frac{\sum_{i=1}^{N} m_i \phi_{ij}}{\sum_{i=1}^{N} m_i \phi_{ij}^2} \quad \text{or} \quad \{\Gamma\} = \frac{[\phi]^T[m]\{1\}}{[\phi]^T[m][\phi]} \tag{7.50}$$

where L_j is the generalized force associated with the jth mode shape.

EXAMPLE 7.2

Consider the two-story building frame shown in Figure E7.1 and determine the participation factors.

SOLUTION

1. Determine the participation factors. First, apply the summation form of Equation 7.50:

$$\Gamma_1 = \frac{L_1}{M_1} = \frac{\sum_{i=1}^{2} m_i \phi_{i1}}{\sum_{i=1}^{2} m_i \phi_{i1}^2} = \frac{m_1 \phi_{11} + m_2 \phi_{21}}{m_1 \phi_{11}^2 + m_2 \phi_{21}^2} = \frac{0.4(0.414) + 0.4(1)}{0.4(0.414)^2 + 0.4(1)^2} = \frac{0.566}{0.469} = 1.207$$

$$\Gamma_2 = \frac{L_2}{M_2} = \frac{\sum_{i=1}^{2} m_i \phi_{i1}}{\sum_{i=1}^{2} m_i \phi_{i1}^2} = \frac{m_1 \phi_{11} + m_2 \phi_{21}}{m_1 \phi_{11}^2 + m_2 \phi_{21}^2} = \frac{0.4(-2.414) + 0.4(1)}{0.4(-2.414)^2 + 0.4(1)^2} = \frac{-0.566}{2.731} = -0.207$$

2. Determine the participation factors by expanding the MATLAB script presented in Example 7.1 (we present the entire script for completeness):

```
clear all % clears any previously defined variables
clc % clears the screen
m = [0.4 0; 0 0.4]; % mass matrix
k = [60 -20; -20 20]; % stiffness matrix
[phi,lam]=eig(k,m); % compute eigenvalues and eigenvectors
omegas=sqrt(lam); % determine and show frequencies from eigenvalues
periods = 2*pi*diag(inv(omegas)); % determine the periods from omegas
% mode shapes by normalizing eigenvectors to get top displ equal to 1
[N_rows, N_cols] = size(phi); % finds N, the size of the matrices
for i = 1:N_cols; % loops over the N modes to normalize them
    norm_phi(:,i) = phi(:,i)./phi(N_rows,i);
end
% determine the participation factors
LT = sum(norm_phi'*m,2) % sum(A,2) returns a vector of sums of each row
MT= sum(norm_phi'*m*norm_phi,2)
par_fac = LT./MT % ./ operation divides each element of two vectors
```

The results of this script are the same as those of part (i):
LT =
  0.5657
 −0.5657

MT =
  0.4686
  2.7314

par_fac =
  1.2071
 −0.2071

Also, in MDOF systems, a modal participation factor can be viewed as providing a measure of the degree to which the jth mode contributes (or participates) to the total dynamic response. Equation 7.50 is essentially the same as Equation 6.32 for the generalized SDOF case, except the use of N eigenvectors rather than a single assumed mode shape. Therefore, we can use the formulations presented in Chapter 6 to obtain the response of a system due to each eigenvector. That is, the original system of N simultaneous differential equations of motion is now transformed into a system of N-independent differential equations, each of which can be solved using the SDOF procedures described in Chapters 3–5. Thus, for linear elastic systems the total displacement can be obtained using superposition of the modal contributions:

$$u_j(t) = \sum_{i=1}^{N} \phi_{ji} \cdot q_i(t) \quad \text{or} \quad \{u(t)\} = \{\phi\}_1 q_1(t) + \{\phi\}_2 q_2(t) + \cdots + \{\phi\}_N q_N(t) \qquad (7.51)$$

where i denotes the level, from 1 to N, and j the range of eigenvectors, from 1 to N. With these displacements, we can compute equivalent lateral forces, which can then be used to conduct a static structural analysis to determine the internal element forces and stresses. This superposition approach produces the entire time history of the structural response (displacements, forces, etc.). However, for design of various elements of the structure, we only need the absolute maximum values of the response. This is the basis for the code-based seismic response history modal analysis procedure introduced in Section 8.7. A more efficient approach is to use the response spectrum analysis described in the next section.

7.1.3 DEFLECTIONS, BASE SHEAR, AND MOMENTS USING PARTICIPATION FACTORS AND RESPONSE SPECTRA

The approach covered in Section 6.1.3 for a generalized SDOF (a single mode of vibration) can be used to solve MDOF systems by treating each mode shape (eigenvector) as an independent generalized SDOF system. After computing the periods for each mode of vibration, T_{nj}, using the eigenvalues, we can determine the maximum dynamic response caused by any dynamic excitation (or force) using the SDOF system analyses discussed in Chapters 3 and 4. Also, we can use the response spectrum analysis method discussed in Chapter 5 to determine the peak response to an earthquake characterized by a design response spectrum with spectral displacement, velocity, and acceleration (D_j, V_j, and A_j, respectively). The values of D_j, V_j, and A_j are obtained as the ordinates of the response spectrum corresponding to the periods T_{nj} associated with the jth mode of vibration, and an appropriate damping ratio. Thus, the solution to the equation of motion associated with the jth mode of vibration based on response spectrum analysis using the spectral displacement, D_j, gives the maximum displacements at the top of the building (maximum principle coordinates) as

$$q_{jo} = \Gamma_j \cdot D_j \qquad (7.52)$$

These maximum displacements can be distributed to the other floors using the normalized eigenvectors, or mode shapes. That is, the maximum displacement value of the ith floor, induced, when only the jth mode shape is excited, is

$$u_{ijo} = \phi_{ij} \cdot q_{jo} = \phi_{ij} \cdot \Gamma_j \cdot D_j \qquad (7.53)$$

The displacement subscript ij can be interpreted as the contribution to the total response at level i due to mode of vibration j. In matrix form, each matrix column represents the floor displacements associated with each mode shape, that is,

$$[u_o] = [\phi][q_o] = [\phi][\Gamma][D] \qquad (7.54)$$

where:
    $[q_o]$ is a diagonal matrix of maximum principle coordinates
    $[\Gamma]$ is a diagonal matrix of participation factors
    $[D]$ is a diagonal matrix of spectral displacements

The displacements can also be written in terms of the spectral accelerations A_j as

$$\left| u_{ijo} \right| = \phi_{ij} \cdot q_{jo} = \phi_{ij} \cdot \Gamma_j \cdot \frac{A_j}{\omega_{nj}^2} \qquad (7.55)$$

The maximum acceleration value of the ith floor, induced when only the jth mode shape is excited, is

$$\left| \ddot{u}_{ijo} \right| = \phi_{ij} \Gamma_j A_j = \phi_{ij} \Gamma_{ij} \omega_{nj}^2 D_j \qquad (7.56)$$

Similar to displacement, this can be represented in matrix form as

$$[\ddot{u}_o] = [\phi][\Gamma][A] = [\phi][\Gamma][\omega_n^2][D] \qquad (7.57)$$

where:
   $[\omega_n^2]$ is a diagonal matrix of the square modal frequencies (also known as the spectral matrix)
   $[A]$ is a diagonal matrix of spectral accelerations

The equivalent static forces associated with these floor accelerations can be obtained using Newton's second law; the product of the associated story mass and maximum story acceleration; that is, the inertial force at level i due to vibration mode j (using Equations 7.56) is

$$f_{ij} = m_i \phi_{ij} \Gamma_j \omega_{nj}^2 D_j \quad \text{or} \quad f_{ij} = m_i \phi_{ij} \Gamma_j A_j \qquad (7.58)$$

Similar to accelerations, this can be represented in matrix form as

$$[f] = [m][\phi][\Gamma]\left[\omega_n^2\right][D] = [m][\phi][\Gamma][A] \qquad (7.59)$$

From static analysis, we can determine the force at the base of the building, which is known as the *base shear*, V_{jb}, for each mode j,

$$V_{jb} = f_{1j} + f_{2j} + f_{3j} + \cdots = \Gamma_j \omega_{nj}^2 D_j \sum_{i=1}^{N} m_i \phi_{ij} = \Gamma_j \omega_{nj}^2 D_j L_j \quad \text{or} \quad V_{jb} = \Gamma_j A_j L_j \qquad (7.60)$$

The base shear in matrix form is

$$\{V_b\} = [f]^T \{1\} = [\Gamma][A][\phi]^T[m]\{1\} = [\Gamma][D]\left[\omega_n^2\right][\phi]^T[m]\{1\} \qquad (7.61)$$

Figure 7.5 shows the equivalent static forces along with the base shear and overturning moment for the jth mode shape. To obtain the overturning moment, we can apply static equilibrium to the free-body diagram.

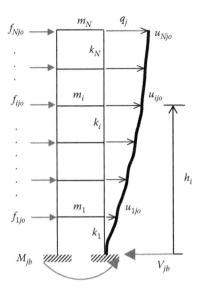

FIGURE 7.5   Maximum dynamic displacements and associated equivalent static forces.

The inertial forces (or lateral story forces) at level i due to vibration mode j can also be written in terms of the base shear by substituting the second of Equations 7.60 as $\Gamma_j A_j = V_{jb}/L_j$ into the second of Equations 7.58:

$$f_{ij} = m_i \phi_{ij} \Gamma_j A_j = m_i \phi_{ij} \frac{V_{jb}}{L_j} \quad \text{or} \quad f_{ij} = \frac{m_j \phi_{ij} V_{jb}}{\sum_{j=1}^{N} m_j \phi_{ji}} \tag{7.62}$$

Also, writing the second of Equations 7.60 in the form of Newton's second law ($F = mA$), we get

$$V_{jb} = M_j^e A_j \tag{7.63}$$

where M_j^e is the *effective mass* given as

$$M_j^e = \Gamma_j L_j \quad \text{or} \quad M_j^e = \frac{L_j^2}{M_j} = \frac{\left(\sum_{i=1}^{N} m_i \phi_{ij}\right)^2}{\sum_{i=1}^{N} m_i \phi_{ij}^2} \tag{7.64}$$

Substituting into Equation 7.63 the definition of the mass in terms of the floor weights, $m_i = w_i/g$ (weight associated with the ith floor divided by the acceleration due to gravity),

$$V_{jb} = \frac{\left(\sum_{i=1}^{N} w_i \phi_{ij}\right)^2}{\sum_{i=1}^{N} w_i \phi_{ij}^2} \frac{A_j}{g} = W_j^e \frac{A_j}{g} \tag{7.65}$$

where W_j^e is the *effective weight* given as

$$W_j^e = \frac{\left(\sum_{i=1}^{N} w_i \phi_{ij}\right)^2}{\sum_{i=1}^{N} w_i \phi_{ij}^2} \tag{7.66}$$

Combining the contribution of all the modes of vibration, we get the *total weight* of the building as

$$W_{total} = \sum_{j=1}^{N} W_j^e \qquad (7.67)$$

This implies that the effective weight of the *j*th mode shape is the fraction of the total weight that participates in the *j*th mode of vibration. The code (ASCE-7) requires including enough modes of vibration in the analysis to obtain an effective weight of at least 90% of the actual building weight, as will be discussed in Section 8.6.

Displacement, acceleration, and lateral story forces resulting from different mode shapes must be combined to acquire the total response, though a simple summation is not used since modal maxima generally occur at different times during the response history. Furthermore, response spectra only provide the values of the modal maxima and not the time at which each value occurs. Therefore, appropriate combination rules have been developed to obtain the total response. These rules are based on random vibration theory to estimate the average maximum response. The most popular rules include the SRSS and CQC methods, the latter being more complicated, but yielding more accurate results for a wider range of mode shapes. The following relationships for the SRSS and CQC are given in terms of generic maximum modal values, R_j, which are intended to represent maximum values due to vibration mode *j* for any response parameter, such as displacement, acceleration, etc.

The total response for the SRSS rule

$$R_{max} \approx \sqrt{\sum_{j=1}^{N} R_j^2} \qquad (7.68)$$

The total response for the CQC rule

$$R_{max} \approx \sqrt{\sum_{i=1}^{N} \sum_{j=1}^{N} R_j \rho_{ij} R_j} \qquad (7.69)$$

where:

$$\rho_{ij} = \frac{8\varsigma^2(1+\beta_{ij})\beta_{ij}^{1.5}}{(1+\beta_{ij}^2)^2 + 4\beta_{ij}\varsigma^2(1+\beta_{ij})^2} \qquad (7.70)$$

ρ_{ij} is the cross-modal coefficient that varies from 0 to 1 (1 for the case of $i = j$), and is generally expressed in terms of the modal frequencies of two distinct mode shapes, $\beta_{ij} = \omega_j/\omega_i$ and damping characteristics, which reduces to the damping ratio given by Equation 3.32 (repeated here for convenience, $\zeta = c/c_{cr}$) when the modal damping is constant for the entire modal spectrum. Also, when the modal frequencies are well separated, this matrix tends to the identity matrix and the CQC rule approaches the SRSS rule results.

To determine an upper bound to both of these rules, we can sum the maximum absolute values as

$$R_{max} \leq \sum_{j=1}^{N} R_j \qquad (7.71)$$

Also, as discussed in Section 3.3, with the maximum dynamic displacement or acceleration, we can conduct a static structural analysis to determine element forces (bending moment, shear force, and axial force), and stresses needed for design; no additional dynamic analysis is necessary. These maximum element force or stress results from each mode shape can be combined using either SRSS or CQC.

EXAMPLE 7.3

Consider the two-story building frame of Figure E7.1 and determine (1) effective masses, (2) peak displacements, (3) maximum equivalent static floor forces, (4) maximum base shear, and (5) maximum overturning moment. Assume beams are rigid, damping of 5%, and that the frame is subjected to a ground acceleration due to an earthquake characterized by the 84.1% design spectrum scaled to 0.25g PGA, the solution is given as follows.

SOLUTION

1. Determine the effective masses. Use Equation 7.64:

$$M_1^e = \frac{L_1^2}{M_1} = \frac{\left(\sum_{i=1}^{2} m_i\phi_{i1}\right)^2}{\sum_{i=1}^{2} m_i\phi_{i1}^2} = \frac{(m_1\phi_{11} + m_2\phi_{21})^2}{m_1\phi_{11}^2 + m_2\phi_{21}^2} = \frac{(0.4(0.414) + 0.4(1))^2}{0.4(0.414)^2 + 0.4(1)^2} = \frac{0.320}{0.469} = 0.683$$

$$M_2^e = \frac{L_2^2}{M_2} = \frac{\left(\sum_{i=1}^{2} m_i\phi_{i1}\right)^2}{\sum_{i=1}^{2} m_i\phi_{i1}^2} = \frac{(m_1\phi_{11} + m_2\phi_{21})^2}{m_1\phi_{11}^2 + m_2\phi_{21}^2} = \frac{(0.4(-2.414) + 0.4(1))^2}{0.4(-2.414)^2 + 0.4(1)^2} = \frac{0.320}{2.731} = 0.117$$

The operation can be performed using MATLAB after determining participation factors, using the same factors, MT and LT; that is,

```
Eff_mass = LT.^2./MT  % .^ operation squares each element of LT vector
```

2. Determine peak (maximum) floor displacements. Determine the maximum displacement at the top of the building, $q_{jo} = \Gamma_j \cdot D_j$, where D_j is due to an 84.1% design spectrum, scaled to 0.25g PGA; first, enter the response spectrum in Figure 5.11 with the following two values for T_n:

   $T_{n1} = 1.16$ s, which is greater than 0.66 s, so the acceleration is given as,
   $A_1 = \eta \cdot 1.8g/T_{n1} = 0.25(1.8\text{ s})(386.4\text{ in/s}^2)/1.16\text{ s} = 149.9\text{ in/s}^2$, and
   $T_{n2} = 0.48$ s, which is greater than 1/8, but less than 0.66 s, so the acceleration is constant,
   $A_2 = \eta \cdot 2.71g = 0.25(2.71)(386.4\text{ in/s}^2) = 261.8\text{ in/s}^2$,

where:
   η is the PGA scale factor for a given earthquake, in this case 0.25
   The maximum spectral displacements are

$$D_1 = \frac{A_1}{\omega_{n1}^2} = \frac{149.9\text{ in/s}^2}{(5.41\text{ rad/s})^2} = 5.12\text{ in}$$

$$D_2 = \frac{A_1}{\omega_{n2}^2} = \frac{261.8\text{ in/s}^2}{(13.1\text{ rad/s})^2} = 1.52\text{ in}$$

Now, determine the maximum floor displacements using Equation 7.53, $u_{ijo} = \phi_{ij} \cdot \Gamma_j \cdot D_j$

$$u_{i1o} = \phi_{i1}\Gamma_1 \cdot D_1 = \begin{Bmatrix} 0.414 \\ 1 \end{Bmatrix}(1.207)(5.12\text{ in}) = \begin{Bmatrix} 2.56 \\ 6.18 \end{Bmatrix}\text{ in}$$

$$u_{i2o} = \phi_{i2}\Gamma_2 \cdot D_2 = \begin{Bmatrix} -2.414 \\ 1 \end{Bmatrix}(-0.207)(1.52\text{ in}) = \begin{Bmatrix} 0.76 \\ -0.32 \end{Bmatrix}\text{ in}$$

Finally, combine these displacements using the SRSS rule (Equation 7.68) to obtain the peak displacements at each level:

$$u_{1\max} = \sqrt{(2.56\,\text{in})^2 + (0.76\,\text{in})^2} = 2.67\,\text{in}$$

$$u_{2\max} = \sqrt{(6.18\,\text{in})^2 + (-0.32\,\text{in})^2} = 6.19\,\text{in}$$

3. Determine maximum equivalent static floor forces. Determine the equivalent static floor forces using Equation 7.58 or 7.59; let us use the matrix form:

$$[f] = [m][\phi][\Gamma][A]$$

$$= \begin{bmatrix} 0.4 & 0 \\ 0 & 0.4 \end{bmatrix} \begin{bmatrix} 0.414 & -2.414 \\ 1 & 1 \end{bmatrix} \begin{bmatrix} 1.207 & 0 \\ 0 & -0.207 \end{bmatrix} \begin{bmatrix} 149.9 & 0 \\ 0 & 261.8 \end{bmatrix}$$

$$= \begin{bmatrix} 0.4(0.414)(1.207)(149.9) & 0.4(-2.414)(-0.207)(261.8) \\ 0.4(1.207)(149.9) & 0.4(-0.207)(261.8) \end{bmatrix}$$

$$= \begin{bmatrix} 30.0 & 52.3 \\ 72.4 & -21.7 \end{bmatrix} \text{kips}$$

Finally, combine these forces using the SRSS rule (Equation 7.68) to obtain the maximum floor forces at each level:

$$f_{1\max} = \sqrt{(30\,\text{kips})^2 + (52.3\,\text{kips})^2} = 60.3\,\text{kips}$$

$$f_{2\max} = \sqrt{(72.4\,\text{kips})^2 + (-21.7\,\text{kips})^2} = 75.5\,\text{kips}$$

4. Determine maximum base shear. Add the equivalent static floor forces for each mode using Equation 7.60 or 7.61; let us use the matrix form again:

$$\{V_b\} = [f]^T\{1\} = \begin{bmatrix} 30.0 & 72.4 \\ 52.3 & -21.7 \end{bmatrix} \begin{Bmatrix} 1 \\ 1 \end{Bmatrix} = \begin{Bmatrix} 30.0 + 72.4 \\ 52.3 - 21.7 \end{Bmatrix} = \begin{Bmatrix} 102.4 \\ 30.6 \end{Bmatrix} \text{kips}$$

Finally, combine these using the SRSS rule (Equation 7.68) to obtain the maximum base shear:

$$V_{b\max} = \sqrt{(102.4\,\text{kips})^2 + (30.6\,\text{kips})^2} = 106.9\,\text{kips}$$

5. Determine maximum overturning moment. Use static equilibrium to obtain the contribution from each mode to the maximum overturning moment:

$$M_{b1} = 30\,\text{kips}(15\,\text{ft}) + 72.4\,\text{kips}(30\,\text{ft}) = 2622\,\text{kip}\cdot\text{ft}$$

$$M_{b2} = 52.3\,\text{kips}(15\,\text{ft}) - 21.7\,\text{kips}(30\,\text{ft}) = 133\,\text{kip}\cdot\text{ft}$$

Finally, combine these using the SRSS rule (Equation 7.68) to obtain the maximum overturning moment:

$$M_{b\max} = \sqrt{(2622\,\text{kip}\cdot\text{ft})^2 + (133\,\text{kip}\cdot\text{ft})^2} = 2625\,\text{kip}\cdot\text{ft}$$

7.2   RESPONSE SPECTRUM ANALYSIS METHOD SUMMARY

The following is a brief step-by-step procedure to estimate the maximum response of a structure subjected to a ground excitation due to an earthquake characterized by a response spectrum scaled to η peak ground acceleration:

1. Determine the mass matrix, $[m]$ from the given floor weights.
2. Determine the stiffness matrix, $[k]$ from the column properties for a shear building.
3. With the stiffness and mass matrices, solve the associated eigenvalue problem for eigenvalues, ω^2, which are used to determine the natural frequencies, ω_n and periods, T_n.
4. After selecting an appropriate damping ratio, determine spectral accelerations (or displacements) using the ordinates of the acceleration (or displacement) response spectrum of the excitation for each of the natural periods.
5. With the eigenvalues, we can also determine the eigenvectors, which are normalized to obtain the modal matrix, $[\phi]$.
6. Obtain the modal participation factors, $\{\Gamma\} = [\phi]^T[m]\{1\}/[\phi]^T[m][\phi]$; (the contribution of each mode shape to the total response).
7. Determine the displacements associated with each mode shape, $[u] = [\phi][\Gamma][D] = [\phi][\Gamma][A][\omega^2]^{-1}$

   where:

   > $[\Gamma]$ = diagonal matrix of participation factors
   > $[D]$ = diagonal matrix of spectral displacements
   > $[A]$ = diagonal matrix of spectral accelerations
   > $[\omega^2]$ = diagonal matrix of squared modal frequencies

8. The resultant maximum displacement at each node is obtained using the SRSS rule (Equation 7.68) for each row vector: $u_{maxi} = (\Sigma u_i^2)^{1/2}$.
9. The matrix of lateral forces at each node is: $[f] = [k][u]$.
10. The resultant maximum lateral force at each node is obtained using SRSS of each row vector: $f_{maxi} = (\Sigma f_i^2)^{1/2}$.
11. The column vector of total base shear forces is: $\{V_b\} = [f]^T\{1\}$
12. The maximum base shear force is obtained using SRSS as: $V_{bmax} = (\Sigma V_i^2)^{1/2}$.
13. The maximum overturning moment is obtained from SRSS using static equilibrium.

EXAMPLE 7.4

Given the three-story building frame shown in Figure E7.2 with damping of 5% and subjected to a ground acceleration due to an earthquake characterized by the 84.1% design spectrum scaled to 0.25g PGA, determine (1) periods, (2) mode shapes, (3) peak displacements, (4) maximum equivalent static floor forces, (5) maximum base shear, and (6) maximum overturning moment. Assume beams are rigid. Each story has a stiffness $k = 326.3$ kips/in (Figure E7.2).

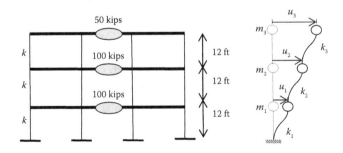

FIGURE E7.2   Building frame schematic (left) and idealized MDOF structural model (right).

SOLUTION

1. Determine the equations of motion, frequencies, periods, and shape vectors. The equations of motion for this case can be determined using D'Alembert's principle by applying horizontal equilibrium to the free-body diagrams of the three masses shown in Figure E7.3. The three equations of motion are

$$+ \rightarrow \Sigma F_x = 0; \quad -m_1\ddot{u}_1 - k_1 u_1 + k_2(u_2 - u_1) = 0 \Rightarrow m_1\ddot{u}_1 + (k_1 + k_2)u_1 - k_2 u_2 = 0$$

$$+ \rightarrow \Sigma F_x = 0; \quad -m_2\ddot{u}_2 - k_2(u_2 - u_1) + k_3(u_3 - u_2) = 0 \Rightarrow m_2\ddot{u}_2 - k_2 u_1 + (k_2 + k_3)u_2 - k_3 u_3 = 0$$

$$+ \rightarrow \Sigma F_x = 0; \quad -m_3\ddot{u}_3 - k_3(u_3 - u_2) = 0 \Rightarrow m_3\ddot{u}_3 - k_3 u_2 + k_3 u_3 = 0$$

These three dependent equations of motion are second-order, linear, and homogeneous differential equations with constant coefficients, and can be rewritten in the matrix form as

$$
\begin{bmatrix} m_1 & 0 & 0 \\ 0 & m_2 & 0 \\ 0 & 0 & m_3 \end{bmatrix}
\begin{Bmatrix} \ddot{u}_1 \\ \ddot{u}_2 \\ \ddot{u}_3 \end{Bmatrix}
+
\begin{bmatrix} k_1 + k_2 & -k_2 & 0 \\ -k_2 & k_2 + k_3 & -k_3 \\ 0 & k_3 & k_3 \end{bmatrix}
\begin{Bmatrix} u_1 \\ u_2 \\ u_3 \end{Bmatrix}
=
\begin{Bmatrix} 0 \\ 0 \\ 0 \end{Bmatrix}
$$

Since $k_1 = k_2 = k_3 = k = 326.3$ kips/in and $m_1 = m_2 = m = 100$ kips/g and $k_3 = 50$ kips/g,

$$
[m] =
\begin{bmatrix} 100\text{ kips} & 0 & 0 \\ 0 & 100\text{ kips} & 0 \\ 0 & 0 & 50\text{ kips} \end{bmatrix}
/386.4\,\text{in/s}^2
$$

$$
[k] =
\begin{bmatrix} 2 & -1 & 0 \\ -1 & 2 & -1 \\ 0 & -1 & 1 \end{bmatrix}
\cdot 326.3\,\text{kips/in}
$$

We use a MATLAB script to solve the eigenvalue problem:

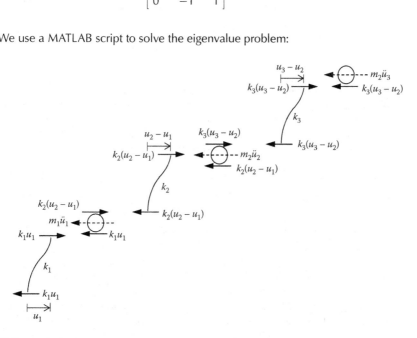

FIGURE E7.3 FBDs for each mass used to determine equations of motion.

```
clear all % clears any previously defined variables
clc % clears the screen
m = [100 0 0; 0 100 0; 0 0 50]/386.4% mass matrix in kips-sec²/in
k = [2 -1 0; -1 2 -1; 0 -1 1]*326.3% stiffness matrix in kips/in
[phi,lam]=eig(k,m); % compute eigenvalues and eigenvectors
omegas=sqrt(lam) % determine and show frequencies from eigenvalues
periods = 2*pi*diag(inv(omegas)) % determine the periods from omegas
% mode shapes by normalizing eigenvectors to get top displ equal to 1
[N_rows, N_cols] = size(phi); % finds N, the size of the matrices
for i = 1:N_cols; % loops over the N modes to normalize them
    norm_phi(:,i) = phi(:,i)./phi(N_rows,i);
end
norm_phi % display normalized modal matrix
```

The results of the script are:

omegas =
    18.3803   0   0
          0 50.2160   0
          0   0   68.5963

periods =
    0.3418
    0.1251
    0.0916

norm_phi =
    0.5000 −1.0000   0.5000
    0.8660   0.0000 −0.8660
    1.0000   1.0000   1.0000

2. With the eigenvalues and eigenvectors, first we follow the same procedure as for the generalized SDOF case using the first mode (following the same process as in Example 6.1). Then, we use matrix operations to complete the entire solution.
   a. Use the first mode, ϕ_{i1}, as the shape vector:

$$\phi_{i1} = \begin{Bmatrix} 0.5 \\ 0.866 \\ 1 \end{Bmatrix}$$

   b. Determine the generalized properties. The building frame can be modeled as a generalized SDOF system assuming that the beams are rigid and only lateral deformations of columns occur. The stiffness of each floor is then obtained by summing the column lateral stiffnesses at each level. The generalized mass, stiffness, and force of the generalized SDOF system are calculated as follows:

Level, i	m_i	k_i	ϕ_{i1}	$\Delta\phi_{i1}$	$m_i \cdot \phi_{i1}$	$m_i \cdot \phi_{i1}^2$	$k_i \cdot (\Delta\phi_{i1})^2$
3	$m/2$	k	1	0.134	$m/2 \cdot (1)$	$m/2 \cdot (1)^2$	$k \cdot (0.134)^2$
2	m	k	0.866	0.366	$m \cdot (0.866)$	$m \cdot (0.866)^2$	$k \cdot (0.366)^2$
1	m	k	0.5	0.5	$m \cdot (0.5)$	$m \cdot (0.5)^2$	$k \cdot (0.5)^2$
		$\sum_{i=1}^{3} =$			1.866m	1.5m	0.402k

Substituting the given values for mass and stiffness, the following properties of the "generalized SDOF," which are the first elements of the modal mass and stiffness matrices, are calculated as follows.

Generalized mass:

$$M_1 = \sum_{i=1}^{3} m_i \cdot \phi_{i1}^2 = 1.5m = 1.5\frac{W}{g} = 1.5\frac{100 \text{ kips}(1000 \text{ lb/kip})}{386.4 \text{ in/s}^2} = 388.2 \text{ lb-s}^2/\text{in}$$

Generalized stiffness:

$$K_1 = \sum_{i=1}^{3} k_i \cdot (\Delta\phi_{i1})^2 = 0.402\,k = 0.402(326.3 \text{ kips/in})(1000 \text{ lb/kip}) = 131{,}144 \text{ lb/in}$$

Generalized force:

$$L_1 = \sum_{i=1}^{3} m_i \cdot \phi_{i1} = 1.866 \cdot m = 1.866\frac{W}{g} = 1.866\frac{100 \text{ kips}(1000 \text{ lb/kip})}{386.4 \text{ in/s}^2} = 482.9 \text{ lb-s}^2/\text{in}$$

c. Determine the natural period, frequency, and participation factor for the first mode. Natural period, which is the same as the MATLAB result in step 1:

$$T_{n1} = 2\pi\sqrt{\frac{M_1}{K_1}} = 2\pi\sqrt{\frac{388.2 \text{ lb-s}^2/\text{in}}{131{,}144 \text{ lb/in}}} = 0.3418 \text{ s}$$

Natural frequency, which is the same as the MATLAB result in step 1:

$$\omega_{n1} = \sqrt{\frac{K_1}{M_1}} = \sqrt{\frac{131{,}144 \text{ lb/in}}{388.2 \text{ lb-s}^2/\text{in}}} = 18.38 \text{ rad/s}$$

Participation factor:

$$\Gamma_1 = \frac{L_1}{M_1} = \frac{482.9 \text{ lb-s}^2/\text{in}}{388.2 \text{ lb-s}^2/\text{in}} = 1.244$$

d. Determine maximum floor displacements. Determine the maximum displacement at the top of the building, $q(t)$ or D_1 due to an 84.1% design spectrum, scaled to 0.25g PGA; first, enter the response spectrum in Figure 5.11 with $T_{n1} = 0.342$ s, which is greater than 1/8 s so the acceleration is constant:

$$A_1 = \eta \cdot 2.71g = 0.25(2.71)(386.4 \text{ in/s}^2) = 261.8 \text{ in/s}^2$$

The maximum spectral displacement is

$$D_1 = \frac{A_1}{\omega_{n1}^2} = \frac{261.8 \text{ in/s}^2}{(18.38 \text{ rad/s})^2} = 0.775 \text{ in}$$

We can now determine the maximum floor displacements using Equation 6.35:

$$u_{io} = \phi_{i1}\Gamma_1 \cdot D_1 = \begin{Bmatrix} 0.5 \\ 0.866 \\ 1 \end{Bmatrix}(1.244)(0.775 \text{ in}) = \begin{Bmatrix} 0.482 \\ 0.835 \\ 0.964 \end{Bmatrix} \text{in}$$

vs. Example problem 1 in Chapter 6, where $u_{jo} = \begin{Bmatrix} 0.311 \\ 0.623 \\ 0.934 \end{Bmatrix}$ in

e. Determine the equivalent static story forces. Determine the equivalent static story forces using Equation 6.37,

$$f_{io} = \Gamma_1 m_i \phi_{i1} A_1$$

$$f_{3o} = \Gamma_1 m_3 \phi_{31} A_1 = 1.244(50 \text{ kips}/g)(1)(0.25 \cdot 2.71 \cdot g) = 42.1 \text{ kips}$$

$$f_{2o} = \Gamma_1 m_2 \phi_{21} A_1 = 1.244(100 \text{ kips}/g)(0.866)(0.25 \cdot 2.71 \cdot g) = 73.0 \text{ kips}$$

$$f_{1o} = \Gamma_1 m_1 \phi_{11} A_1 = 1.244(100 \text{ kips}/g)(0.5)(0.25 \cdot 2.71 \cdot g) = 42.1 \text{ kips}$$

f. Determine base shear and overturning moment. Determine the base shear force using Equation 6.40,

$$V_{\max b} = \sum_{i=1}^{3} f_{io} = 42.1 \text{ kips} + 73.0 \text{ kips} + 42.1 \text{ kips} = 157.3 \text{ kips}$$

Determine the overturning moment using Equation 6.41,

$$M_{\max b} = \sum_{j=1}^{3} h_j f_{oj} = 42.1 \text{ kips}(12 \text{ ft}) + 73.0 \text{ kips}(24 \text{ ft}) + 42.1 \text{ kips}(36 \text{ ft}) = 3775 \text{ kip} \cdot \text{ft}$$

As in Chapter 6, we can obtain the internal story shear force and moment at each level by drawing shear force and bending moment diagrams considering the building as a cantilever beam. The diagrams also compare these results to those of the analysis of Example 6.1, which are given in parentheses (Figure E7.4).

At this point, we could follow steps 2-a to 2-f for each of the remaining mode shapes and then apply one of the combination rules to obtain the total response. Alternatively, we could use the matrix operation process listed in Section 7.3, beginning with step 6 to perform the analysis of all the modes simultaneously, see next step.

3. Use the matrix operations listed in Section 7.3:

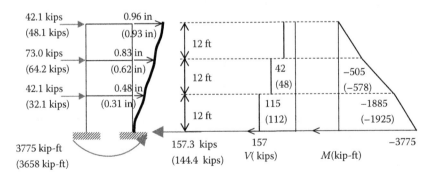

FIGURE E7.4   Lateral forces, internal shear, and internal moment diagram for mode 1.

a. Obtain the modal participation factors, $\{\Gamma\} = [\phi]^T[m]\{1\}/[\phi]^T[m][\phi]$: First, get the modal matrix, $[M] = [\phi]^T[m][\phi]$,

$$[M] = \begin{bmatrix} 0.5 & 0.866 & 1 \\ -1 & 0 & 1 \\ 0.5 & -0.866 & 1 \end{bmatrix} \begin{bmatrix} 100 & 0 & 0 \\ 0 & 100 & 0 \\ 0 & 0 & 50 \end{bmatrix} \begin{bmatrix} 0.5 & -1 & 0.5 \\ 0.866 & 0 & -0.866 \\ 1 & 1 & 1 \end{bmatrix} / 386.4 \text{ in/s}^2$$

$$[M] = \begin{bmatrix} 0.388 & 0 & 0 \\ 0 & 0.388 & 0 \\ 0 & 0 & 0.388 \end{bmatrix} \text{lb-s}^2/\text{in}$$

And, the modal force, $[L] = [\phi]^T[m]\{1\}$,

$$[L] = \begin{bmatrix} 0.5 & 0.866 & 1 \\ -1 & 0 & 1 \\ 0.5 & -0.866 & 1 \end{bmatrix} \begin{bmatrix} 100 & 0 & 0 \\ 0 & 100 & 0 \\ 0 & 0 & 50 \end{bmatrix} \begin{Bmatrix} 1 \\ 1 \\ 1 \end{Bmatrix} / 386.4 \text{ in/s}^2 = \begin{Bmatrix} 0.483 \\ -0.129 \\ 0.0347 \end{Bmatrix} \text{lb-s}^2/\text{in}$$

Now, get the participation factors, $\{\Gamma\} = [\phi]^T[m]\{1\}/[\phi]^T[m][\phi]$,

$$\{\Gamma\} = [M]^{-1}\{L\} = \begin{bmatrix} 2.576 & 0 & 0 \\ 0 & 2.576 & 0 \\ 0 & 0 & 2.576 \end{bmatrix} \text{in/lb-s}^2 \cdot \begin{Bmatrix} 0.483 \\ -0.129 \\ 0.0347 \end{Bmatrix} \text{lb-s}^2/\text{in} = \begin{Bmatrix} 1.244 \\ -0.333 \\ 0.0893 \end{Bmatrix}$$

b. Determine the displacements associated with each mode shape, $[u] = [\phi][\Gamma][A][\omega^2]^{-1}$. First, assemble the participation factors and square of the frequencies into diagonal matrices:
Diagonal matrix of participation factors:

$$[\Gamma] = \begin{bmatrix} 1.244 & 0 & 0 \\ 0 & -0.333 & 0 \\ 0 & 0 & 0.0893 \end{bmatrix}$$

Diagonal matrix of squared modal frequencies, which is equal to the eigenvalues:

$$[\omega_n^2] = \begin{bmatrix} 337.8 & 0 & 0 \\ 0 & 2521.6 & 0 \\ 0 & 0 & 4705.5 \end{bmatrix} \text{rad}^2/\text{s}^2$$

To obtain the diagonal matrix of accelerations, we first enter the response spectrum in Figure 5.11 with the following three values for T_n:
- $T_{n1} = 0.342$ s, which is greater than 1/8 s so the acceleration is constant,
  $A_1 = \eta \cdot 2.71g = 0.25(2.71)(386.4 \text{ in/s}^2) = 261.8 \text{ in/s}^2$;
- $T_{n2} = 0.1251$ s, which is also greater than 1/8 s so the acceleration is constant,
  $A_2 = \eta \cdot 2.71g = 0.25(2.71)(386.4 \text{ in/s}^2) = 261.8 \text{ in/s}^2$; and
- $T_{n3} = 0.0916$ s, which is less than 1/8 s so the acceleration is given as,
  $A_3 = \eta \cdot 11.7(T_{n3})^{0.704}g = 0.25(11.7)(0.0916)^{0.704}(386.4 \text{ in/s}^2) = 210.1 \text{ in/s}^2$.

Thus, the diagonal matrix of spectral accelerations is given as

$$[A] = \begin{bmatrix} 261.8 & 0 & 0 \\ 0 & 261.8 & 0 \\ 0 & 0 & 210.1 \end{bmatrix} \text{in/s}^2$$

The displacements then, $[u] = [\phi][\Gamma][A][\omega^2]^{-1}$

$$[u] = \begin{bmatrix} 0.5 & -1 & 0.5 \\ 0.866 & 0 & -0.866 \\ 1 & 1 & 1 \end{bmatrix} \begin{bmatrix} 1.244 & 0 & 0 \\ 0 & -0.333 & 0 \\ 0 & 0 & 0.0893 \end{bmatrix}$$

$$* \begin{bmatrix} 261.8 & 0 & 0 \\ 0 & 261.8 & 0 \\ 0 & 0 & 210.1 \end{bmatrix} \begin{bmatrix} 337.8 & 0 & 0 \\ 0 & 2521.6 & 0 \\ 0 & 0 & 4705.5 \end{bmatrix}^{-1}$$

$$= \begin{bmatrix} 0.482 & 0.0346 & 0.0020 \\ 0.835 & 0 & -0.0035 \\ 0.964 & -0.0346 & 0.0040 \end{bmatrix} \text{in}$$

c. The resultant maximum displacement at each node is obtained from the SRSS rule (Equation 7.68) of each row vector: $u_{maxi} = (\Sigma u_i^2)^{1/2}$:

$$u_{1max} = \sqrt{(0.482 \text{ in})^2 + (0.0346 \text{ in})^2 + (0.002 \text{ in})^2} = 0.483 \text{ in}$$

$$u_{2max} = \sqrt{(0.835 \text{ in})^2 + (-0.0035 \text{ in})^2} = 0.835 \text{ in}$$

$$u_{3max} = \sqrt{(0.964 \text{ in})^2 + (0.0346 \text{ in})^2 + (0.004 \text{ in})^2} = 0.965 \text{ in}$$

d. The matrix of lateral forces at each node is, $[f] = [k][u]$:

$$[f] = \begin{bmatrix} 2 & -1 & 0 \\ -1 & 2 & -1 \\ 0 & -1 & 1 \end{bmatrix} \cdot 326.3 \text{ kip/in} \cdot \begin{bmatrix} 0.482 & 0.0346 & 0.0020 \\ 0.835 & 0 & -0.0035 \\ 0.964 & -0.0346 & 0.0040 \end{bmatrix} \text{in}$$

$$= \begin{bmatrix} 42.1 & 22.6 & 2.4 \\ 73.0 & 0 & -4.2 \\ 42.1 & -11.3 & 2.4 \end{bmatrix} \text{kips}$$

These are shown in Figure E7.5.

e. The resultant maximum lateral force at each node is obtained from the SRSS rule (Equation 7.68) of each row vector: $f_{imax} = (\Sigma f_i^2)^{1/2}$,

$$f_{1max} = \sqrt{(42.1 \text{ kips})^2 + (22.6 \text{ kips})^2 + (2.4 \text{ kips})^2} = 47.9 \text{ kips}$$

$$f_{2max} = \sqrt{(73 \text{ kips})^2 + (-4.2 \text{ kips})^2} = 73.1 \text{ kips}$$

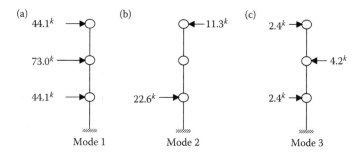

FIGURE E7.5 Resulting lateral forces at each node for a given mode of vibration: (a) Mode 1, (b) Mode 2, and (c) Mode 3.

$$f_{3\max} = \sqrt{(42.1\,\text{kips})^2 + (-11.3\,\text{kips})^2 + (2.4\,\text{kips})^2} = 43.7\ \text{kips}$$

f. The column vector of total base shear forces is: $\{V_b\} = [f]^T\{1\}$,

$$[V_b] = \begin{bmatrix} 42.1 & 73.0 & 42.1 \\ 22.6 & 0 & -11.3 \\ 2.4 & -4.3 & 2.4 \end{bmatrix}\text{kips} \cdot \begin{Bmatrix} 1 \\ 1 \\ 1 \end{Bmatrix} = \begin{Bmatrix} 157.6 \\ 11.3 \\ 0.7 \end{Bmatrix}\text{kips}$$

Alternatively, we can obtain all the story shear forces using static equilibrium at each level, for each mode shape:

$$\{V_1\} = \begin{Bmatrix} 42.1 + 73 + 42.1 \\ 42.1 + 73 \\ 42.1 \end{Bmatrix}\text{kips} = \begin{Bmatrix} 157.6 \\ 115.2 \\ 42.1 \end{Bmatrix}\text{kips}$$

$$\{V_2\} = \begin{Bmatrix} -11.3 + 22.6 \\ -11.3 \\ -11.3 \end{Bmatrix}\text{kips} = \begin{Bmatrix} 11.3 \\ -11.3 \\ -11.3 \end{Bmatrix}\text{kips}$$

$$\{V_3\} = \begin{Bmatrix} 2.4 - 4.3 + 2.4 \\ 2.4 - 4.3 \\ 2.4 \end{Bmatrix}\text{kips} = \begin{Bmatrix} 0.7 \\ -1.8 \\ 2.4 \end{Bmatrix}\text{kips}$$

g. The maximum base shear force is obtained from SRSS as: $V_{b\max} = (\Sigma V_i^2)^{1/2}$,

$$V_{b\max} = \sqrt{(157.6\,\text{kips})^2 + (11.3\,\text{kips})^2 + (0.7\,\text{kips})^2} = 158\ \text{kips}$$

The maximum story shear forces can be obtained using the SRSS rule at each level for all the mode shape story shears obtained in step 3-f:

$$\{V_{\max}\} = \begin{Bmatrix} \sqrt{(157.6\,\text{kips})^2 + (11.3\,\text{kips})^2 + (0.7\,\text{kips})^2} \\ \sqrt{(115.2\,\text{kips})^2 + (-11.3\,\text{kips})^2 + (-1.8\,\text{kips})^2} \\ \sqrt{(42.1\,\text{kips})^2 + (-11.3\,\text{kips})^2 + (2.4\,\text{kips})^2} \end{Bmatrix} = \begin{Bmatrix} 158 \\ 116 \\ 44 \end{Bmatrix}\text{kips}$$

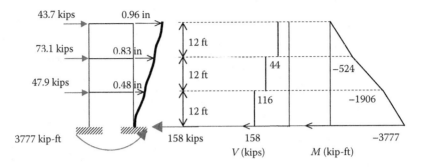

FIGURE E7.6   Lateral forces, internal shears, and internal moment diagrams.

h. The maximum overturning moment is obtained from SRSS using static equilibrium. We first obtain the story overturning moments using static equilibrium at each level, for each mode shape:

$$\{M_1\} = \left\{\begin{array}{c} 42.1(36)+73(24)+42.1(12) \\ 42.1(24)+73(12) \\ 42.1(12) \end{array}\right\} \text{kips}\cdot\text{ft} = \left\{\begin{array}{c} 3774 \\ 1887 \\ 505 \end{array}\right\} \text{kips}\cdot\text{ft}$$

$$\{M_2\} = \left\{\begin{array}{c} -11.3(36)+22.6(12) \\ -11.3(24) \\ -11.3(12) \end{array}\right\} \text{kips}\cdot\text{ft} = \left\{\begin{array}{c} -135 \\ -271 \\ -135 \end{array}\right\} \text{kips}\cdot\text{ft}$$

$$\{M_3\} = \left\{\begin{array}{c} 2.4(36)-4.3(24)+2.4(12) \\ 2.4(24)-4.3(12) \\ 2.4(12) \end{array}\right\} \text{kips}\cdot\text{ft} = \left\{\begin{array}{c} 13 \\ 7 \\ 29 \end{array}\right\} \text{kips}\cdot\text{ft}$$

The maximum story overturning moments can now be obtained using the SRSS rule:

$$\{M_{max}\} = \left\{\begin{array}{c} \sqrt{(3774)^2 +(135)^2 +(13)^2} \\ \sqrt{(1887)^2 +(271)^2 +(7)^2} \\ \sqrt{(505)^2 +(135)^2 +(29)^2} \end{array}\right\} \text{kips}\cdot\text{ft} = \left\{\begin{array}{c} 3777 \\ 1906 \\ 524 \end{array}\right\} \text{kips}\cdot\text{ft}$$

Figure E7.6 summarizes the results of the maximum response obtained:
4. We can write a MATLAB script to perform all the operation as follows:

```
clear all % clears any previously defined variables
clc % clears the screen
g = 386.4; % acceleration due to gravity in in/s²
pA = 0.25* g; % peak ground acceleration
m = [100 0 0; 0 100 0; 0 0 50]/386.4 % mass matrix in kip·s²/in
k = [2 −1 0; −1 2 −1; 0 −1 1]*326.3 % stiffness matrix in kips/in
[phi,lam]=eig(k,m); % compute eigenvalues, lam and eigenvectors, phi
omegas=sqrt(lam) % determine and show frequencies from eigenvalues
periods = 2*pi*diag(inv(omegas)) % determine the periods from omegas
% mode shapes by normalizing eigenvectors to get top displ equal to 1
[N_rows, N_cols] = size(phi); % finds N, the size of the matrices
for i = 1:N_cols; % loops over the N modes to normalize them
    norm_phi(:,i) = phi(:,i)./phi(N_rows,i);
end
norm_phi % display normalized modal matrix
% The spectral accelerations from an applicable response spectrum
```

```
SA1 = pA*2.71; % THIS CHANGES WITH THE PERIOD
SA2 = pA*2.71; % THIS CHANGES WITH THE PERIOD
SA3 = pA*11.7*(periods(3))^0.704; % THIS CHANGES WITH THE PERIOD
SA = [SA1,SA2,SA3]
% Top floor spectral displacements
SD = SA*inv(lam)
% determine the participation factors
LT = sum(norm_phi'* m,2) % sum(A,2) returns a vector of sums of each row
MT= sum(norm_phi'* m*norm_phi,2)
par_fac = LT./MT % ./ operation divides each element of the two vectors
% Maximum displacements at each level for each mode shape
for i = 1:N_rows;
    for j = 1:N_cols;
        ui_max(i,j) = SD(j)*norm_phi(i,j)*par_fac(j);
    end
end
ui_max % display displs.; each column corresponds to a mode shape
% Use the SRSS rule to get max displacements at each level
u_maxsrss=sqrt(sum((ui_max).^2,2)) %.^ operation squares each element
% Floor forces at each level for each mode shape
f=k*ui_max
% Use SRSS method to get total max forces at each level
f_maxsrss=sqrt(sum((f).^2,2))
% Base shear for each mode shape and max base shear using SRSS rule
V=(sum(f,1))
V_maxsrss=sqrt(sum((V).^2,2))
% Overturning moment for each mode shape and max moment using SRSS
heights=[12;24;36];
OTM = f'*heights
OMT_maxsrss=sqrt(sum((OTM).^2,1))
```

The results of the script are:

$m =$
```
  0.2588      0      0
       0  0.2588      0
       0      0  0.1294
```

$k =$
```
  652.6000 -326.3000        0
 -326.3000 652.6000 -326.3000
        0  -326.3000  326.3000
```

omegas =
```
  18.3803      0      0
       0  50.2160      0
       0      0  68.5963
```

periods =
```
  0.3418
  0.1251
  0.0916
```

norm_phi =
```
  0.5000  -1.0000   0.5000
  0.8660   0.0000  -0.8660
  1.0000   1.0000   1.0000
```

SA =
```
  261.7860 261.7860 210.0515
```

SD =
  0.7749   0.1038   0.0446

LT =
  0.4829
 −0.1294
  0.0347

MT =
  0.3882
  0.3882
  0.3882

par_fac =
   1.2440
  −0.3333
   0.0893

ui_max =
   0.4820   0.0346   0.0020
   0.8348   −0.0000   −0.0035
   0.9640   −0.0346   0.0040

u_maxsrss =
  0.4832
  0.8348
  0.9646

f =
   42.1411   22.5833   2.4277
   72.9905   −0.0000   −4.2049
   42.1411   −11.2917   2.4277

f_maxsrss =
  47.8724
  73.1115
  43.6951

V =
  157.2726   11.2917   0.6505

V_maxsrss =
  157.6788

OTM =
  1.0e+003 *
   3.7745
  −0.1355
   0.0156

OMT_maxsrss =
  3.7770e+003

PROBLEMS

7.1 Using the definition of stiffness and mass influence coefficients, formulate the equation of motion for the frame shown below. Assume the beams are rigid.

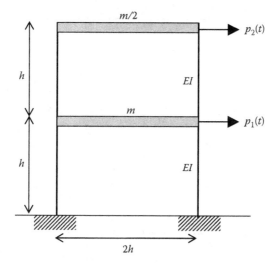

7.2 Formulate the equation of motion (in matrix form) using D'Alembert's principle for the given three-story building, which has rigid beams and flexible steel ($E = 29,000$ ksi) columns with total moment of inertial for each floor as shown.

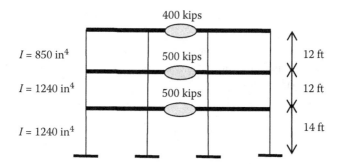

7.3 Formulate the equation of motion (in matrix form) using D'Alembert's principle for the given two-story building, which has rigid beams and flexible steel ($E = 29,000$ ksi) columns with total moment of inertial for each floor as shown.

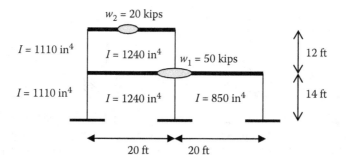

7.4 Formulate the equation of motion (in matrix form) using D'Alembert's principle for the given three-story building, which has rigid beams and flexible steel ($E = 29,000$ ksi) column sections as shown.

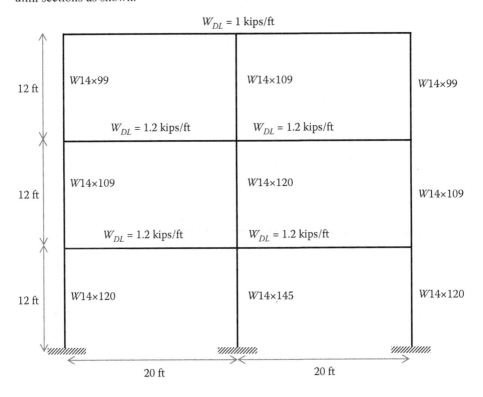

7.5 Formulate the equation of motion (in matrix form) using D'Alembert's principle for the given two-story building, and determine the natural frequencies, periods, normalized modal matrix, and participation factors.

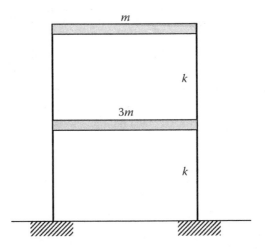

7.6 The following two-story building is supported with steel frames spaced at 15 ft on center, and has the floor load shown plus wall load of 20 psf. Formulate the equation of motion (in matrix form) using D'Alembert's principle and determine the natural frequencies, periods, normalized modal matrix, and participation factors using MATLAB.

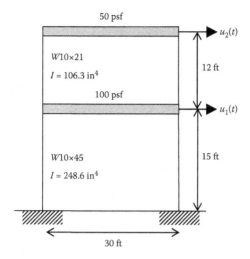

7.7 Given the following mode shapes and frequencies, compute the participation factors for a 3-story building with floor weights of 120 kips for the first floor, and 80 kips for the second floor and roof.

$$[\phi] = \begin{bmatrix} 1.68 & -1.208 & -0.714 \\ 1.22 & 0.704 & 1.697 \\ 0.572 & 1.385 & -0.984 \end{bmatrix} \quad \text{and} \quad \{\omega\} = \begin{Bmatrix} 8.77 \\ 25.18 \\ 48.13 \end{Bmatrix} \text{rad/s}$$

Note that $[\phi]^T[M][\phi] = [I]$, so $\{\Gamma\} = [\phi]^T[M]\{1\}$.

7.8 Use an MDOF analysis for the following building frame to determine the story displacements, story forces, base shear, and overturning moment due to an earthquake characterized by an 84.1% design spectrum, scaled to 0.25g PGA. Assume rigid beams and damping of 5%.

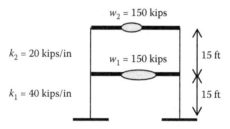

7.9 Given a five-story building frame subjected to a ground acceleration due to a 0.25g PGA earthquake characterized by the 84.1% design spectrum, use an MDOF analysis to determine (i) peak displacements, (ii) maximum base shear, and (iii) maximum overturning moments. $k = 326.3$ kips/in, $w_5 = 50$ kips, and the weights on the other four floors equal 100 kips. Assume rigid beams and damping of 5%.

REFERENCES

Chopra, A. K., *Dynamics of Structures: Theory and Applications to Earthquake Engineering*, 4th edition, Prentice-Hall, Upper Saddle River, NJ, 2012.

Villaverde, R., *Fundamental Concepts of Earthquake Engineering*, CRC Press, Boca Raton, FL, 2009.

8 Seismic Code Provisions

After reading this chapter, you will be able to:

1. Compute seismic load combinations
2. Determine the redundancy factor
3. Explain the overstrength factor
4. Perform calculations used to determine earthquake loads using the equivalent lateral force procedure (ELFP)
5. Perform calculations used to determine earthquake loads using modal response spectrum (MRS) analysis
6. Describe seismic response history procedures

Design of structures must be based on a minimum standard of care to safeguard the safety, health, and welfare of the public. Municipal, state, or federal governments concerned with the safety of the public have established building codes used to control the construction of structures within their jurisdiction. These design codes specify among other things the design loads. While building codes may vary from city to city, most municipalities rely on regional codes and design standards for the specification of loads. These codes and standards include general building codes, such as the International Building Code (IBC) and the ASCE/SEI (2010) "Minimum Design Loads for Buildings and Other Structures" standard, from here on referred to as ASCE-7. These codes and standards are developed by various organizations and present the best opinion of those organizations as to what constitutes good practice. In this chapter, we provide an overview of structural design philosophies and explain how seismic loads are incorporated as part of structural design. The primary focus of this chapter is to describe the parameters associated with determining earthquake loads in accordance with ASCE-7 and to demonstrate the procedures for determining earthquake loads using the ELFP, seismic response history analysis, and MRS analysis.

8.1 STRUCTURAL DESIGN PHILOSOPHIES

ASCE-7 and IBC adhere to two general design philosophies: working stress design also known as allowable stress design (ASD) and limit states design also known as the load and resistance factor design (LRFD); the primary objective of both design philosophies is to obtain structures that will not fail under an extreme loading condition. Failure of a structure is defined as the state where a loading condition exceeds a limit state. Thus, a limit state is a condition beyond which a structure, or structural component, ceases to fulfill the function for which it was designed. There are two limit states of interest, structural failure and serviceability:

1. A structural failure (or strength) limit state is the more critical of the two because it is concerned with safety and it relates to the maximum load carrying capacity of a structure.
2. A serviceability limit state relates to structural performance under normal service conditions.

Both design philosophies focus on the ultimate limit state and allow the designer some freedom of judgment in serviceability. The main concern from a strength limit state standpoint is ensuring that the capacity (resistance, R) of the structure exceeds the demand (effect of loads, Q). Since it is economically impracticable to design 100% safe structures, we must assess the risk of failure, or the reliability of the structure. To assess the reliability of a structure, we must understand how the uncertainties on material properties, geometry, and material durability relate to the load uncertainties; not all the extremes of loads can occur simultaneously on the structure, or, at least, it is very unlikely. In ASD, all uncertainties are accounted using a single quantity, the factor of safety (*FS*), while the LRFD uses a rational approach to obtain a more uniform assessment of the reliability of a structure. Thus, the main difference between ASD and LRFD is that LRFD rationally accounts for the statistical variability in the loads (and materials), and is based on reliability analysis to obtain a probability-based assessment of structural safety.

8.1.1   ALLOWABLE STRESS DESIGN

In ASD, the risk assessment of failure is made on the basis of an allowable *FS*, which was learned from previous experiences for the system in question under its anticipated environment. The *FS* is defined as the ratio of nominal values of resistance R and the sum of load effects ΣQ. That is, to ensure safety in structural design,

$$FS \geq \frac{R}{\Sigma Q} \tag{8.1}$$

It is important to note that both sides of the inequality must be evaluated for the same conditions. For example, load effect that produces tensile stresses should be compared with the tensile strength of the member in question. The strength of a member is material dependent; a properly designed system of members must ensure vertical and lateral load-resisting system integrity by providing ductility, energy dissipation, and avoiding progressive collapse by providing redundancy and preserving the load path—all while maintaining deformations within the prescribed limits.

The stress produced by the loads, ΣQ, is determined once the structural form has been selected using combinations of loads that can be reasonably expected to act on the structure. That is, not all possible loads will act concurrently; thus, ASCE-7 (Section 2.4) specifies the following load combinations:

1. D
2. $D + L$
3. $D + (L_r$ or S or $R)$
4. $D + 0.75 L + 0.75 (L_r$ or S or $R)$
5. $D + (0.6 W$ or $0.7 E)$
6a. $D + 0.75 L + 0.75(0.6 W) + 0.75 (L_r$ or S or $R)$
6b. $D + 0.75 L + 0.75(0.7 E) + 0.75 S$
7. $0.6 D + 0.6 W$
8. $0.6 D + 0.7 E$

where:

D represents the dead loads, which are loads that have constant magnitude, remain in one position, and are usually known to a high degree of certainty. They include the weight of various structural elements and weight of objects permanently attached to the structure, such as walls, roofs, ceilings, and equipment

L represents the live loads, which are those loads that vary in magnitude and position with time. They are caused by the structure being occupied, used, and maintained. In general, live loads are induced by gravity, and can be stationary or transient, which may still be treated as a static load, unless applied rapidly, in which case, the dynamic effect creates additional inertial load called impact

L_r represents the roof live loads, which are loads produced by repair/maintenance workers, their equipment and materials, or during the life of the structure by movable objects such as planters. There is a limit to the amount of load that can be realistically placed upon roofs because most roofs are sloped

S represents the snow load, which depends on ground snow weight, the building's general shape, and roof geometry (flat roofs are subjected to higher loads compared to sloped roofs)

R represents the rain load, which depends on the functionality of the drainage facilities, and only a concern when rainwater accumulates faster than it runs off, ponding. According to ASCE-7, elements should be designed to support all rainwater that accumulates when primary drains are blocked and the water rises above the inlet of the secondary drainage system

W represents the wind load, which is a lateral load caused by wind pressure resulting from the kinetic energy of moving air as it strikes an object in its path

E represents the earthquake load, which is the main subject of this book

These loads are actions that result from the weight of all building materials, occupants and their possessions, environmental effects, and are categorized based on their character and duration. The history of each of these loads can be represented schematically as shown in Figure 8.1.

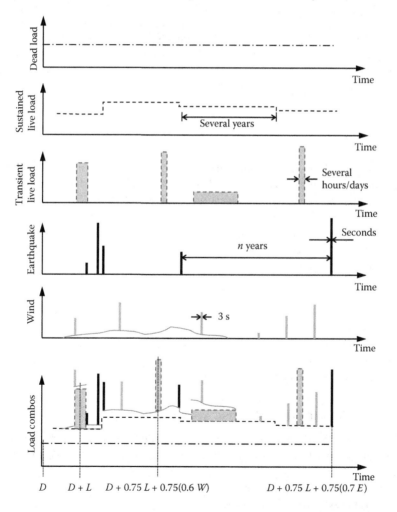

FIGURE 8.1 Schematic history of loads and their superposition to obtain the load combinations.

This figure shows that the dead load remains constant over its entire history, the live load can be divided into a sustained load (one that may have variations at different points in its history, such as building partitions) and a transient load (one that changes frequently, both in duration and magnitude), the earthquake load is based on short events as was discussed in earlier chapters, and a wind load that includes sustained intensity and wind gusts. As shown in the figure, the largest magnitudes of each of these loads are not likely to occur simultaneously, which is why various load combinations are needed to represent potential critical loadings at different points during the life of a structure.

Gravity-induced loads on any floor system are assumed to act as a uniform pressure on the floor slab. We are only concerned with the structural system; so to analyze the effects of these pressure-like forces on each structural member, we need to rationally distribute them to the supporting members. To accomplish this, we use the concept of the tributary area, the area of the slab that is carried by a particular structural member.

8.1.2 Load and Resistance Factor Design

In LRFD, the risk assessment of failure uses probability theory to account for the variability of resistance and load effects, providing a more uniform measure of reliability. The resistance side, R, is multiplied by statistically based resistance factors ϕ that take into account uncertainties in material and geometric properties, and equations that predict strength; while the load effects side is multiplied by statistically based load factors γ_i that take into account uncertainties in the magnitude and position of loads. Because of uncertainties, both Q and R are assumed to be random variables, and their behavior is characterized by a mean, standard deviation, and frequency distribution. The safety criterion is then stated as

$$\phi R \geq \sum \gamma_i Q_i \tag{8.2}$$

Again, both sides of the inequality must be evaluated for the same conditions. The left-hand side of this inequality is the available strength, while the right-hand side is the required strength, which is computed using the following load combinations (ASCE-7, Section 2.3):

1. $1.4 D$
2. $1.2 D + 1.6 L + 0.5 (L_r$ or S or $R)$
3. $1.2 D + 1.6 (L_r$ or S or $R) + (L$ or $0.5 W)$
4. $1.2 D + W + L + 0.5 (L_r$ or S or $R)$
5. $1.2 D + E + L + 0.2 S$
6. $0.9 D + W$
7. $0.9 D + E$

where all quantities were previously defined. Also, in this design philosophy, unfactored loads are used for serviceability requirements because safety is not usually of concern for serviceability.

8.1.3 Allowable Seismic Force-Resisting System

The resistance side of the inequalities for both design philosophies is related to the force-resisting system. Allowable seismic force-resisting systems are listed in ASCE-7, Table 12.2-1, and include:

a. Bearing wall systems (in concrete, masonry, and timber)
b. Building frame systems (in steel, concrete, masonry, and timber)
c. Moment-resisting frame systems (in steel and concrete)

    d. Dual systems with special moment frames (in steel and concrete)
    e. Dual systems with intermediate moment frames (in steel and concrete)
    f. Ordinary shearwall frame interactive systems (in concrete)
    g. Cantilever column systems (in steel, concrete, and timber)
    h. Steel systems not detained for seismic resistance

Some of these systems are not permitted in seismic-resistant design; and those permitted are subject to building height and other limitations. These limitations depend on the various seismic design categories (SDCs, discussed later in Section 8.4.2, and ASCE-7, Section 11.6), which are classifications of building requirements based on a desired performance that depends on a building's occupancy, the effects of probable ground-shaking intensity at the building site, and structural irregularities.

8.2 OVERVIEW OF SEISMIC DESIGN CODES

Section 1.3 of the book provides a brief history of the development of mitigation strategies for the effects of seismic hazards, primarily shaking effects. Since 2000, a new model-building code has been in use, the aforementioned IBC, which refers the user to the ASCE-7 standard for seismic-loading specifications. ASCE-7 incorporates a substantial portion of the National Earthquake Hazard Reduction Program (NEHRP) provisions for seismic design. The explicit objective of these seismic provisions is to ensure that strength and deflection (drift) requirements are met while preserving the integrity of the load path throughout the structural system during a seismic event. And, while not explicitly stated, the objectives of the IBC seismic provisions are as follows:

1. Resist a minor earthquake with no damage.
2. Resist a moderate earthquake with some nonstructural damage, but no structural damage.
3. Resist a major earthquake with some structural damage, but no collapse.

In all cases, damage depends on: intensity and duration of ground shaking, building configuration (irregular buildings are more susceptible to damage), lateral force-resisting system type (e.g., shearwalls, braced or unbraced frames), building material (e.g., concrete, steel, wood, or masonry), and quality of construction. Note that IBC or ASCE-7 provides no guidance for protection against earth movement, earth slides, liquefaction, or direct fault displacement. The ASCE-7 seismic loading-related specifications are provided in 13 chapters, the titles of which are as follows:

Chapter 11: *Seismic Design Criteria*
Chapter 12: *Seismic Design Requirements for Building Structures*
Chapter 13: Seismic Design Requirements for Nonstructural Components
Chapter 14: Material-Specific Seismic Design and Detailing Requirements
Chapter 15: Seismic Design Requirements for Nonbuilding Structures
Chapter 16: *Seismic Response History Procedures*
Chapter 17: Seismic Design Requirements for Seismically Isolated Structures
Chapter 18: Seismic Design Requirements for Structures with Damping Systems
Chapter 19: Soil Structure Interaction for Seismic Design
Chapter 20: *Site Classification Procedure for Seismic Design*
Chapter 21: *Site-Specific Ground Motion Procedures for Seismic Design*
Chapter 22: *Seismic Ground Motion and Long-Period Transition Maps*
Chapter 23: Seismic Design Reference Documents

    This chapter only covers some of the seismic code provisions, mainly procedures described in Chapters 11, 12, and 16 of ASCE-7.

8.3   SEISMIC LOAD COMBINATIONS

The earthquake load E is computed using one of the methods allowed by the code that will be presented later in this chapter. This load characterizes vertical and lateral load effects, which are combined following the provisions presented in ASCE-7, Section 12.4. All members of the vertical and lateral force-resisting systems must be designed using the load combinations presented earlier; when seismic effects govern, load combinations 5, 6b, and 8 are used in ASD, while load combinations 5 and 7 are used in LRFD. The values of E used in load combinations of both design philosophies are based on a combination of horizontal, E_h, and vertical, E_v, components, and are given by the following relationships:

1. When effects of seismic and gravity loads are additive,

$$E = E_h + E_v = \rho Q_E + 0.2 S_{DS} D \tag{8.3}$$

2. When gravity loads counteract the seismic effects,

$$E = E_h - E_v = \rho Q_E - 0.2 S_{DS} D \tag{8.4}$$

For cases where structural overstrength should be included:

1. When effects of seismic and gravity loads are additive,

$$E = E_m = E_h + E_v = \Omega_o Q_E + 0.2 S_{DS} D \tag{8.5}$$

2. When gravity loads counteract the seismic effects,

$$E = E_m = E_h - E_v = \Omega_o Q_E - 0.2 S_{DS} D \tag{8.6}$$

where:
   E_v is not a peak value; rather it recognizes that the peak values of the horizontal and vertical seismic forces are unlikely to occur simultaneously. Also, E_v need not be included when $S_{DS} \leq 0.125g$
   ρ is the redundancy factor and is discussed in Section 8.3.1
   Q_E is the effect of horizontal seismic forces (exhibited as internal forces in the axial, shear, and flexure members), and will be discussed in Sections 8.6–8.8
   D is the dead load previously defined
   S_{DS} is the design spectral acceleration for short periods, and will be discussed in Section 8.5
   Ω_o is the overstrength factor discussed in Section 8.3.2

Regarding the direction of the design seismic loads, ASCE-7 (Section 12.5) specifies that E be applied in directions that produce the largest effect on structural members. The code does, however, provide guidelines to satisfy this requirement. For example, design seismic forces are permitted to be applied separately and independently in each of two orthogonal directions for structures assigned to SDC B (discussed in Section 8.4.2) and regular structures in all other categories.

EXAMPLE 8.1

A column of a steel special concentrically braced frame in a single story medical office building (SDC = D) supports an axial dead load, $D = 35$ kips, and horizontal seismic load effect, $Q_E = \pm 15$ kips. Given $S_{DS} = 1.25$, an overstrength factor, $\Omega_o = 2.5$, and a redundancy factor,

$\rho = 1.3$, determine the maximum and minimum axial forces in the column using LRFD load combinations. Repeat the problem accounting for overstrength.

SOLUTION

1. Determine the maximum axial load. Determine the maximum earthquake load, E, using Equation 8.3:

$$E = E_h + E_v = \rho Q_E + 0.2S_{DS}D = 1.3(\pm 15 \text{ kips}) + 0.2(1.25)(35 \text{ kips}) = \pm 19.5 \text{ kips} + 8.8 \text{ kips}$$

Thus, taking the positive quantity in the first term, the maximum earthquake load is

$$E = E_h + E_v = 19.5 \text{ kips} + 8.8 \text{ kips} = 23.8 \text{ kips}$$

Now, determine the maximum axial load using load combination 5:

$$5) \quad 1.2D + E + L + 0.2S = 1.2(35 \text{ kips}) + 28.3 \text{ kips} = 70.3 \text{ kips}$$

2. Determine the minimum axial load. Determine the minimum earthquake load, E, using Equation 8.4:

$$E = E_h - E_v = \rho Q_E - 0.2S_{DS}D = 1.3(\pm 15 \text{ kips}) - 0.2(1.25)(35 \text{ kips}) = \pm 19.5 \text{ kips} - 8.8 \text{ kips}$$

Thus, taking the negative quantity in the first term, the minimum earthquake load is

$$E = E_h - E_v = -19.5 \text{ kips} - 8.8 \text{ kips} = -28.3 \text{ kips}$$

Now, determine the minimum axial load using load combination 7:

$$7) \quad 0.9D + E = 0.9(35 \text{ kips}) - 28.3 \text{ kips} = 3.2 \text{ kips}$$

3. Determine the maximum axial load including *overstrength*. Determine the maximum earthquake load, E_m, using Equation 8.5:

$$E_m = E_h + E_v = \Omega_o Q_E + 0.2S_{DS}D = 2.5(\pm 15 \text{ kips}) + 0.2(1.25)(35 \text{ kips}) = \pm 37.5 \text{ kips} + 8.8 \text{ kips}$$

Thus, taking the positive quantity in the first term, the maximum earthquake load is

$$E_m = E_h + E_v = 37.5 \text{ kips} + 8.8 \text{ kips} = 46.3 \text{ kips}$$

Now, determine the maximum axial load using load combination 5:

$$5) \quad 1.2D + E_m + L + 0.2S = 1.2(35 \text{ kips}) + 46.3 \text{ kips} = 88.3 \text{ kips}$$

4. Determine the minimum axial load including *overstrength*. Determine the minimum earthquake load, E_m, using Equation 8.6:

$$E_m = E_h - E_v = \Omega_o Q_E - 0.2S_{DS}D = 1.3(\pm 15 \text{ kips}) - 0.2(1.25)(35 \text{ kips}) = \pm 19.5 \text{ kips} - 8.8 \text{ kips}$$

Thus, taking the negative quantity in the first term, the minimum earthquake load is

$$E_m = E_h - E_v = -37.5 \text{ kips} - 8.8 \text{ kips} = -46.3 \text{ kips}$$

Now, determine the minimum axial load using load combination 7:

7) $0.9D + E_m = 0.9(35 \text{ kips}) - 46.3 \text{ kips} = -14.8 \text{ kips}$

8.3.1 REDUNDANCY FACTOR, ρ

Redundancy is a characteristic in which multiple paths of resisting lateral loads are provided. ρ is intended to encourage system redundancy; when redundancy is lacking, ρ in Equations 8.3 and 8.4 can result in a penalty, or increase in the seismic load of up to 30%. Also, ρ is assigned to the seismic force-resisting system in each of two orthogonal directions for all structures. For the following conditions, ρ can be taken as 1.0, and thus there is no penalty:

1. All structures in SDC *B* or *C*.
2. Calculations of the drift and *P*-delta (*P*-Δ) effects.
3. For the design of nonstructural components and nonbuilding structures.
4. For the design of members, connections, collector elements, and splices, when load combinations with the overstrength factor are required.
5. Diaphragm design forces specified in ASCE-7, Section 12.10.1.1.
6. Design of structures with damping systems designed in accordance with ASCE-7, Chapter 18.

The redundancy factor should be taken as 1.3 (a 30% penalty) for structures assigned to SDC *D*, *E*, and *F*. Unless one of the following two conditions is met, whereby ρ is permitted to be taken as 1.0:

1. Stories resisting forces larger than 35% of the base shear in any given direction that complies with Table 12.3-3 in ASCE-7.
2. Structures that are regular in plan at all levels and are provided with at least two bays of seismic force-resisting perimeter framing on each side of the structure, in each orthogonal direction, with each story resisting more than 35% of the base shear. For masonry, concrete, or steel plate shearwalls, the number of bays shall be calculated as the length of shearwall divided by the story height. For light-framed construction shearwalls, the number of bays shall be calculated as two times the length of shearwall divided by the story height.

EXAMPLE 8.2

A single-story building shown in Figure E8.1 with an average roof height of 12 ft is constructed with a flexible roof diaphragm assigned to SDC *D*. The base shear was determined to be 100 kips,

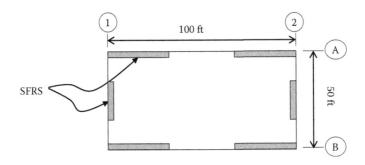

FIGURE E8.1 Single-story building plan view.

which can be resolved into total forces along wall lines 1 and 2 of 50 kips each. Determine the redundancy factor for each of three lateral force-resisting systems consisting of (1) 18-foot long concrete shearwalls, (2) single-steel ordinary concentrically braced frames, and (3) double-steel special moment frames.

SOLUTION

1. 18-foot long concrete shearwalls.
   Determine the number of bays using one of the following relationships: Light-frame construction number of bays is given as

$$\# bays = 2L_w/h_{sx}$$

where:
   L_w is the length of the shearwall
   h_{sx} is the story height

For all other cases, such as concrete,

$$\# bays = L_w/h_{sx}$$

Thus, the number of bays for the concrete shearwalls is

$$\# bays_{csw} = L_w/h_{sx} = 18 \text{ ft}/12 \text{ ft} = 1.5 < 2 \text{ bays minimum}$$

Therefore, $\rho = 1.3$ (no redundancy)

2. Single-steel ordinary concentrically braced frames.

$$\# bays_{SCBF} = 1 < 2 \text{ bays minimum}$$

Therefore, $\rho = 1.3$

3. Double-steel special moment frames.

$$\# bays_{SSMF} = 2 = 2 \text{ bays minimum}$$

Therefore, $\rho = 1.0$

8.3.2 OVERSTRENGTH FACTOR Ω_o AND OTHER SEISMIC FORCE-RESISTING SYSTEM PARAMETERS

To ensure safety of the overall system, a complete lateral load path must be provided. A proper load path is guaranteed by providing sufficient ductility of the system, which is achieved by properly designing and detailing the overall system. Detailing refers to special connection/component designs to integrate the overall lateral force-resisting system in order to transfer the seismic loads from their point of origin to the foundation of the structural system. Ductility of a system is mobilized when seismic-resistant structures experience inelastic deformations during an earthquake. As shown in Section 5.2, ductility greatly reduces the seismic response of the overall structure mainly because of two reasons: (a) yielding of the structure leads to inelastic deformations that change the stiffness of the structure, which leads to the lengthening of the effective period, reducing the required strength demand; and (b) substantial energy dissipation (hysteretic damping) from inelastic behavior.

The effect of ductility is taken into account with the response modification coefficient, R, which reduces the lateral seismic force to strength level design forces. As discussed in Section 5.2, we use elastic response spectra to obtain inelastic ones, which based on ASCE-7 specifications can be

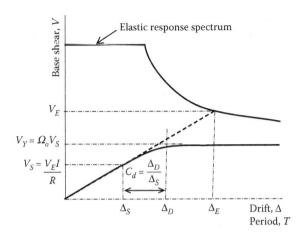

FIGURE 8.2 Inelastic force–deformation curve.

accomplished by dividing elastic spectra by R. The response modification coefficient, R, is given as the ratio of the lateral force developed in a linear elastic structure, V_E, to the prescribed design lateral force, V_S, and represents an adjustment of elastic force to design load level for the entire structural system in order to reach the required ductility demand; see Figure 8.2. Values of R are listed in ASCE-7, Table 12.2-1, for various seismic force-resisting systems. Table 12.2-1 in ASCE 7 provides values for two additional parameters, the overstrength factor Ω_o and the deflection amplification factor C_d. Values of R vary from 1 to 8, Ω_o vary from 1.25 to 3, and C_d vary from 1.25 to 6; for example, both steel and reinforced concrete special moment frames have $R = 8$, $\Omega_o = 3$, and $C_d = 5.5$. These three parameters are related and collectively characterize the inelastic behavior of structures. Figure 8.2 illustrates the relationship between the parameters.

When a component does not reach the necessary level of ductility, the seismic forces can be greatly underestimated. In the methods used to obtain seismic loads discussed later in this chapter, it is assumed that all components have the same level of ductility, which is an extremely rare condition. Some components (like bolted connections) have much lower levels of ductility as compared to other components of the lateral force-resisting system, such as shearwalls and beams. The overstrength factor, Ω_o, attempts to oversize low-level-ductile (or nonyielding) portions of the structure so that first yield failure occurs in the more ductile elements, avoiding sudden brittle failures during a seismic event. The overstrength factor, Ω_o, is given as the ratio of the actual lateral force developed in a structure, V_Y, to the calculated design lateral force, V_S, and represents an adjustment of design force to the actual load level for some components that cannot reach the required ductility demand; see Figure 8.2.

Deflections are typically calculated using elastic structural analyses, and since excessive deformation can lead to instabilities of the overall system, it is important to estimate deflections accurately. Rather than conducting a full inelastic structural analysis, the deflection amplification factor C_d magnifies the deflection of the structure based on an elastic analysis. These amplified lateral deflections (or drifts) must be limited to prescribed values. The deflection amplification factor C_d represents the ratio of the design drift, Δ_D, to the drift under design forces, Δ_S, and relates the elastic drift to the inelastic drift; see Figure 8.2. It is important to note that the specifications require that elastic deflections be calculated at the strength level; that is, using load combinations, which implies a strength limit state and not a serviceability limit state.

EXAMPLE 8.3

Use ASCE-7, Table 12.2-1, to determine seismic force-resisting system parameters and limitations for a steel special moment frame (SSMF) depicted in Figure E8.2. Also, list an advantage and a disadvantage for SSMF systems.

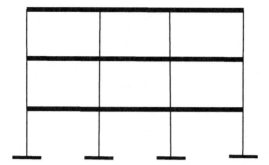

FIGURE E8.2 Special steel moment frame schematic.

SOLUTION

1. Determine seismic force-resisting system parameters. From ASCE-7, Table 12.2-1, the SSMF system is part of group C, C.1. The values of design coefficients and factors are as follows:

   Response modification coefficient, $R = 8$
   The overstrength factor $\Omega_o = 3$
   The deflection amplification factor $C_d = 5.5$
2. Identify the limitations for the seismic force-resisting system. From ASCE-7, Table 12.2-1, group C, C.1, structural system limitations are available according to SDC as

SDC	B	C	D	E	F
Restriction	NL	NL	NL	NL	NL

For each SDC of the SSMF, NL = not limited is listed, which implies that the system is permitted everywhere and has no height limits.
3. Describe an advantage and a disadvantage for SSMF systems.

   Advantage: because of the high R value, the base shear is relatively low.
   Disadvantage: because of the high C_d value, the drifts are relatively high.

8.4 OVERVIEW OF SEISMIC LOAD ANALYSIS PROCEDURES

As discussed in Section 8.3, earthquakes cause vertical and horizontal accelerations, which result in vertical and horizontal forces on structures. Vertical forces are generally smaller than lateral forces and are considered only in overturning analyses. The focus of this section is on determining the effect of horizontal seismic forces, Q_E, in seismic load combinations shown in Equations 8.3–8.6. The methods discussed in this section are used to conduct a complete structural analysis of horizontal forces. After determining a site-specific response spectrum and selecting an allowable seismic force-resisting system, the process entails, calculating the base shear, distributing this base shear force to various levels of the structure, and using structural analysis to determine lateral deflections as well as internal forces and moments generated in each of the structural elements.

8.4.1 DESIGN RESPONSE SPECTRUM

The key step in obtaining lateral forces using any method is the development of appropriate design response spectra. ASCE-7's approach is similar to that followed by Newmark and Hall as described in Chapter 5. ASCE-7 uses two spectral response accelerations (S_S and S_1) that are based on a maximum considered earthquake (MCE) ground motion (as described in Section 2.4).

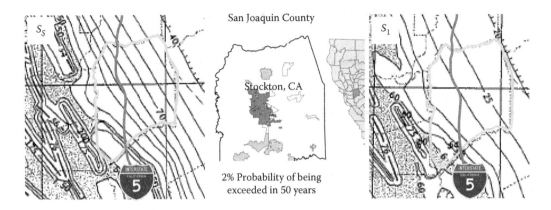

FIGURE 8.3 Maximum considered ground motion intensity in San Joaquin County, California (as percent of acceleration due to gravity).

S_S and S_1 are presented in map form in ASCE-7, Figures 22-1–22-6, assuming 5% damping and a site underlaid by rock, see Figure 8.3, for the contour maps in San Joaquin County, California. S_1 represents acceleration for a period of 1.0 s, while S_S represents acceleration for a period of 0.2 s (short period). These maps are difficult to read, especially in cases where contour lines are closely spaced. A more accurate and convenient approach is to use the USGS Seismic Hazards Mapping Application, available at: http://earthquake.usgs.gov/designmaps/us/application.php, as shown in Figure 8.4. This application generates detailed reports that provide a site-specific response spectrum; see Example 8.4.

As indicated in Chapter 5, Figure 5.13, the response spectrum is affected by local soil conditions. Since S_S and S_1 are based on rock subsurface conditions, they must be adjusted for other

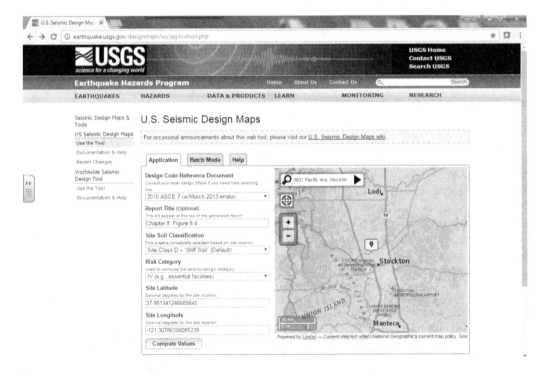

FIGURE 8.4 USGS seismic hazard mapping application input screen. (Courtesy of USGS.)

types of soil conditions. ASCE-7, Chapter 20, classifies a site as Site Class A (hard rock), B (rock), C (very dense soil and soft rock), D (stiff soil), E (soft soil), or F (soil) based on soil shear wave velocity, v_s; if v_s is not known, ASCE-7 allows the use of standard penetration resistance or undrained shear strength values. This information is typically included as part of a geotechnical engineering report. If no soil properties are known, ASCE-7 allows the use of Site Class D as a conservative assumption.

The response spectrum associated with MCE S_S and S_1 can be adjusted to include site class effects, S_{MS} and S_{M1}; the results are given as

$$S_{MS} = F_a S_S \quad \text{and} \quad S_{M1} = F_v S_1 \tag{8.7}$$

where:

F_a is a site coefficient per ASCE-7, Table 11.4-1, that depends on soil type and ranges in values from 0.8 for rock to 2.5 for soft soil

F_v is a site coefficient per ASCE-7, Table 11.4-2, that also depends on soil type and ranges in values from 0.8 for rock to 3.5 for soft soil

S_{MS} is the mapped risk-targeted maximum considered earthquake (MCE_R) spectral response acceleration parameter at short periods

S_{M1} is the mapped MCE_R spectral response acceleration parameter at a period of 1 s

Also, the values F_a and F_v are smaller for large spectral accelerations, which is intended to capture the nonlinear behavior of soils that prevent ground motion amplifications.

The response spectrum associated with S_{MS} and S_{M1} (based on MCE_R ground motion) is intended to produce the most severe earthquake effects. MCE ground motion corresponds to a 2% probability of being exceeded in a 50-year period (~2%/50), which corresponds to an approximate return period, or a recurrence interval of 2500 years, as obtained in Section 2.4. However, the nominal design level is based on a maximum design earthquake (MDE) ground motion, which corresponds to a 10% probability of being exceeded in a 50-year period, or approximately a 500-year recurrence interval (~10%/50). ASCE-7 accounts for two-thirds of the MCE ground motion spectrum accelerations, S_{MS} and S_{M1}, in order to reduce the MCE ground motion from 2%/50 to 10%/50 hazard level; the results are given as

$$S_{DS} = 2/3\, S_{MS} \quad \text{and} \quad S_{D1} = 2/3\, S_{M1} \tag{8.8}$$

where:

S_{DS} is the design, 5% damped, spectral response acceleration parameter at short periods

S_{D1} is the design, 5% damped, spectral response acceleration parameter at a period of 1 s

The design response spectrum associated with S_{DS} and S_{D1} can be obtained using ASCE-7, Section 11.4.5. Draw the design spectral response acceleration, S_a, as a piecewise continuous function, with four segments as shown in Figure 8.5 or algebraically as follows:

For

$$T \leq T_o, \quad S_a = S_{DS}\left(0.4 + 0.6\frac{T}{T_o}\right)$$

where the transition periods T_o and T_s are:

$$T_o = 0.2\, T_s$$

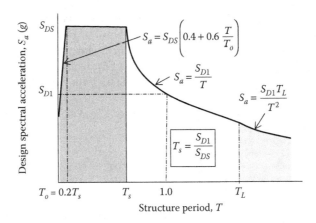

FIGURE 8.5 Elastic design response spectrum.

$T_s = S_{D1}/S_{DS}$, which represents the transition period from acceleration- to velocity-controlled segments of the design spectrum (see Figure 5.8).

T is the fundamental period of the structure.

Note that for $T = 0$, the spectral acceleration, S_a, is approximately equal to $0.4S_{DS}$, which represents the design-level peak ground acceleration.

For $T_o < T \le T_s$, $S_a = S_{DS}$
For $T_s < T \le T_L$, $S_a = (S_{D1}/T)$

and

For $T > T_L$, $S_a = \dfrac{S_{D1}T_L}{T^2}$

The long-period transition period, T_L, is obtained from ASCE-7, Figure 22-12, which is only applicable to tall buildings, or for the sloshing of fluids in tanks. In addition, T_L represents the transition period from velocity- to displacement-controlled portions of the design spectrum (see Figure 5.8).

EXAMPLE 8.4

Determine the spectral response acceleration parameters and draw the response spectrum for a site located on the campus of University of the Pacific in Stockton, California, which will be used for the construction of an office building. The site (soil) is class *D*.

SOLUTION

1. Determine the maximum considered spectral accelerations for short and 1 s periods, S_S and S_1. From the seismic intensity maps in ASCE-7, Chapter 22, S_S and S_1 (as fractions of acceleration due to gravity) are estimated from Figure 8.3,

$$S_S = 0.85$$

$$S_1 = 0.30$$

2. Determine the site coefficients for the given site class. For Site Class *D*, using ASCE-7, Tables 11.4-1 and 11.4-2, identify the site coefficients.

$F_a = 1.16$ using linear interpolation of values in Tables 11.4-1

$F_v = 1.80$ using Tables 11.4-2

3. Determine the maximum considered spectral accelerations, adjusted for site effects, for short and 1 s periods, S_{MS} and S_{M1},

$$S_{MS} = F_a S_S = 1.16(0.85) = 0.896$$
$$S_{M1} = F_v S_1 = 1.80(0.30) = 0.54$$

4. Determine the design spectral accelerations for short and 1 s periods, S_{DS} and S_{D1},

$$S_{DS} = 2/3 S_{MS} = 2/3(0.896) = 0.65$$
$$S_{D1} = 2/3 S_{M1} = 2/3(0.54) = 0.36$$

5. Draw the design response spectrum. The design response spectrum, S_a, is a piecewise continuous function with four segments:

$$\text{For } T \leq T_o, \quad S_a = S_{DS}\left(0.4 + 0.6\frac{T}{T_o}\right) = 0.65\left(0.4 + 0.6\frac{T}{0.11\,\text{s}}\right) = 0.26 + 3.5\,T$$

where the transition periods T_o and T_s are:

$$T_o = 0.2 T_s = 0.11$$

$$T_s = \frac{S_{D1}}{S_{DS}} = \frac{0.36}{0.65} = 0.55\,\text{s}$$

For $T_o < T \leq T_s$, $S_a = S_{DS} = 0.65$
For $T_s < T \leq T_L$, $S_a = (S_{D1}/T) = (0.36/T)$

and

For $T > T_L$, $S_a = (S_{D1}T_L/T^2) = (0.36 T_L/T^2)$

The long-period transition period T_L obtained from ASCE-7, Figure 22-12, is equal to 8 s. Also, notice that for $T = 0$, the spectral acceleration is approximately equal to 0.26, which represents the design-level peak ground acceleration. Figure E8.3 shows the entire spectrum.

The sample report for this example from the USGS Seismic Hazards Mapping Application at http://earthquake.usgs.gov/designmaps/us/application.php also provides the response spectrum (Figure E8.4).

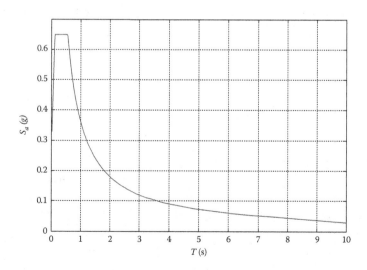

FIGURE E8.3 Acceleration response spectrum at the University of the Pacific, Site Class D.

≋USGS Design Maps Summary Report

User–Specified Input

Report Title Chapter 8, Figure 8.4
Fri October 7, 2016 19:32:21 UTC

Building Code Reference Document ASCE 7-10 Standard
(which utilizes USGS hazard data available in 2008)

Site Coordinates 37.98134°N, 121.30786°W

Site Soil Classification Site Class D – "Stiff Soil"

Risk Category I/II/III

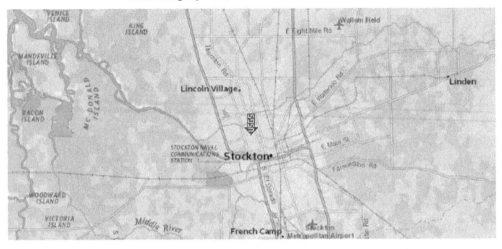

USGS–Provided Output

$S_S =$ 0.896 g $S_{MS} =$ 1.023 g $S_{DS} =$ 0.682 g

$S_1 =$ 0.333 g $S_{M1} =$ 0.578 g $S_{D1} =$ 0.385 g

For information on how the SS and S1 values above have been calculated from probabilistic (risk-targeted) and deterministic ground motions in the direction of maximum horizontal response, please return to the application and select the "2009 NEHRP" building code reference document.

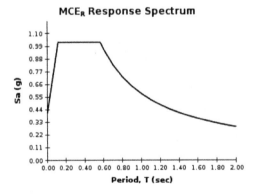

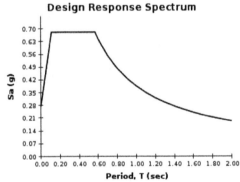

For PGA_M, T_L, C_{RS}, and C_{R1} values, please view the detailed report.

Although this information is a product of the U.S. Geological Survey, we provide no warranty, expressed or implied, as to the accuracy of the data contained therein. This tool is not a substitute for technical subject-matter knowledge.

FIGURE E8.4 USGS seismic hazard mapping application output.

8.4.2 PERMITTED LATERAL ANALYSIS PROCEDURES

SDCs are not only used to determine the allowable seismic force-resisting system and their building height limit as discussed in Section 8.1.3, but also to determine the permitted lateral analysis procedure, as well as restrictions on buildings with irregularities, seismic detailing requirements, and requirements for nonstructural components. SDCs classify building systems into categories A through F based on their desired performance depending on a building's occupancy, the effects of probable ground-shaking intensity at the building site, and structural irregularities. The various SDCs and hazard-level descriptions are:

A → Very low seismic risk
B → Low seismic risk
C → Moderate seismic risk
D → Moderate-to-high seismic risk
E → High seismic risk for structures in Occupancy Categories I, II, and III near an active fault
F → High seismic risk for structures in Occupancy Category IV near an active fault

To determine the SDC for a structure using Table 8.1 (the most severe case of S_{DS}, S_{D1}, or S_1 governs), we must determine the soil site class using ASCE-7, Chapter 20, the MCE spectral response accelerations (S_S and S_1), the design spectral response accelerations (S_{DS} and S_{D1}) as described in Section 8.4.1, and finally the occupancy category using ASCE-7, Table 1.5-1 (discussed in the following paragraphs), or use the USGS application with results provided in the detailed report.

ASCE-7 also classifies buildings based on the occupancy category and importance factor for the building system. Occupancy categories are assigned to various buildings and other structures based on their use and nature of occupancy (see ASCE-7, Table 1.5-1), and they are intended to increase the safety level of essential facilities and those that pose a significant risk to human life. The nominal design level is Occupancy Category II, which corresponds to an MDE ground motion (10%/50). The higher risk levels correspond to the probabilities listed in Table 8.2. As this table shows, essential and critical structures are designed for an MCE ground motion (2%/50), which is the level given by the mapped spectrum parameters in ASCE-7, Chapter 22. Thus, the importance factors listed in Table 8.2 effectively adjust the design response spectrum for various hazard levels.

The importance factors indirectly address the objectives of IBC seismic provisions discussed in Section 8.2, which are related to the performance of systems for various levels of earthquake shaking, that is,

1. Resist a minor earthquake with no damage.
2. Resist a moderate earthquake with some nonstructural damage, but no structural damage.
3. Resist a major earthquake with some structural damage, but no collapse.

TABLE 8.1

Seismic Design Categories; Summary of ASCE-7, Section 11.6

Based on S_{DS} (Table 11.6-1)	Based on S_{D1} (Table 11.6-2)	I or II or III	IV
$S_{DS} < 0.167$	$S_{D1} < 0.067$	A	A
$0.167 \leq S_{DS} < 0.33$	$0.067 \leq S_{D1} < 0.133$	B	C
$0.33 \leq S_{DS} < 0.50$	$0.133 \leq S_{D1} < 0.20$	C	D
$S_{DS} \geq 0.50$	$S_{D1} \geq 0.20$	D	D
	$S_1 \geq 0.75$	E	F

TABLE 8.2

Occupancy Categories and Importance Factors

Occupancy Category	Nature of Occupancy	Return Period (Years)	Importance Factor, I_e
I	Low hazard structures (10% probability of exceedance in 50 years)	475	1.00
II	Standard occupancy, other than I, III, or IV (10% probability of exceedance in 50 years)	475	1.00
III	Assembly structures (5% probability of exceedance in 50 years)	975	1.25
IV	Essential or critical structures (2% probability of exceedance in 50 years)	2475	1.50

Occupancy Categories I and II provide the minimum level of protection and are intended to address the risk of structural failure and loss of life, but not to limit structural damage, whereas Occupancy Categories III and IV provide a higher level of protection against loss of life and property by limiting the damage to a minimum in order to provide a continued function of facilities during and after a seismic event.

The analysis procedures allowed by the code include ASCE-7, Section 12.8, ELFP as well as two dynamic methods: ASCE-7, Section 12.9, MRS analysis, and ASCE-7, Chapter 16, seismic response history procedures, all of which will be discussed in this chapter. The dynamic methods are permitted for all building systems in all six SDCs. The ELFP is permitted for all building systems in SDCs A, B, and C, and is allowed for most cases in SDCs D, E, and F, except buildings exceeding 160 ft in height with $T > 3.5\ T_s$ and those with $T < 3.5\ T_s$ and horizontal irregularities 1a or 1b, or vertical irregularities 1a, 1b, 2, or 3 (as noted in ASCE-7, Table 12.6-1). Even for the exempt cases, the ELFP is used in scaling the results of the dynamic methods.

Building irregularities are discussed in ASCE-7, Section 12.3.2, and are defined as horizontal (ASCE-7, Table 12.3-1) or vertical (ASCE-7, Table 12.3-2). Horizontal irregularities include: (1a) torsional, (1b) extreme torsional, (2) reentrant corners, (3) diaphragm discontinuities, (4) out-of-plane offsets, and (5) nonparallel resisting systems. Vertical irregularities include: (1a) stiffness-soft story, (1b) stiffness-extreme soft story, (2) weight (mass), (3) vertical geometric, (4) In-plane discontinuity in vertical lateral force-resisting element, (5a) discontinuity in lateral strength–weak story, and (5b) discontinuity in lateral strength–extreme weak story.

8.5 EARTHQUAKE LOADS BASED ON ASCE-7 EQUIVALENT LATERAL FORCE PROCEDURE

As discussed in Section 8.4.2, earthquake loads can be obtained using an equivalent static method (as presented here) or a full dynamic analysis approach presented in the following sections. The main purpose of any seismic analysis is to ensure that strength and lateral deflection (drift) requirements are satisfied. ASCE-7 ELFP is similar to the generalized SDOF system analysis for shear buildings presented in Chapter 6. The process involves determining the natural period, the base shear, equivalent static story forces, story drifts, and overturning moment.

The ASCE-7 ELFP analysis steps are as follows (corresponding ASCE-7 sections are given in parenthesis):

1. Compute seismic weight W using details presented in Section 3.1.2 and repeated here for convenience (ASCE-7, Section 12.7.2).

$$W = DL + 0.25 \, StL + \text{larger of } PL \text{ or } 10 - \text{psf} + WPE + 0.2SL + L_aL \qquad (3.8)$$

where:

      DL is the dead load of the structural system that is tributary to each floor

      StL is the storage load, which is a live load in areas used for storage, such as a warehouse

      PL is the partition load, when applicable

      WPE is the operating weight of permanent equipment

      SL is the flat roof snow load when it exceeds 30-psf, regardless of roof slope

      LaL is the landscape loads associated with roof and balcony gardens

2. Compute the natural period of the structure, T, following ASCE-7, Section 12.8.2. T is the calculated fundamental (first mode) period of a structure in the direction under consideration. It can be determined using the structural characteristics of the lateral force-resisting elements as introduced in Section 3.1.1, or the generalized SDOF properties covered in Chapter 6, or a full eigenanalysis covered in Chapter 7. When one of these methods is used to determine T, it is subject to lower and upper limits as follows:

$$T = C_u T_a \quad \text{if} \quad T > C_u T_a$$

$$\text{Use } T \quad \text{if} \quad T_a \leq T \leq C_u T_a$$

$$T = T_a \quad \text{if} \quad T < T_a$$

where:

      C_u is the coefficient for the upper limit on the calculated period from ASCE-7, Table 12.8-1

      T_a is the approximate fundamental period (ASCE-7, Section 12.8.2.1)

Alternatively, the code permits the use of T_a, calculated using the details presented in Section 3.1.4, and repeated here for convenience,

$$T_a = C_t h_n^x \qquad (8.9)$$

where:

      h_n is the height in feet above the base to the highest level of the structure

      C_t and x are determined from Table 3-2 (or ASCE-7, Table 12.8-2)

T_a is based on the measured response of buildings in high seismic regions; its value can be adjusted to get a more accurate approximation of the actual period by including local seismicity using C_u to obtain the upper bound period, $C_u T_a$. This is based on the best-fit of the measured response.

3. Computer seismic base shear (ASCE-7, Section 12.8.1). The base shear is caused by the inertial force (mass times acceleration) from the earthquake motion, $F = ma = W(a/g)$, where g is the acceleration due to gravity. ASCE-7 modifies this expression to obtain an equivalent static lateral force acting at the base of the structure, the *base shear*, V:

$$V = C_s W \qquad (8.10)$$

where:

    W is the seismic weight computed in step 1

    C_s is the seismic response coefficient, expressed as a fraction of g (ASCE-7, Section 12.8.1.1)

$$C_s = \frac{S_{DS}}{R/I_e} \qquad (8.11)$$

where:

    S_{DS} is the design spectral response acceleration in the short period range, computed using the first of Equation 8.8 (ASCE-7, Section 11.4.4)

    R is the response modification factor in ASCE-7, Table 12.2-1

    I_e is the importance factor in Table 8.2 (ASCE-7, Section 11.5.1)

    The value of C_s need not exceed

$$C_s = \frac{S_{D1}}{T(R/I_e)} \quad \text{for} \quad T \le T_L \quad \text{or} \quad C_s = \frac{S_{D1}T_L}{T^2(R/I_e)} \quad \text{for} \quad T > T_L \qquad (8.12)$$

Collectively, these three relationships for C_s represent the inelastic design response spectrum (Equation 8.11 is the constant acceleration portion and Equations 8.12 represents the constant velocity and displacement portions, respectively), obtained by dividing the elastic design response spectrum (shown in Figure 8.5) by the factor R/I_e. Figure 8.6 provides an illustration of the resulting inelastic design response spectrum.

    Also,

$$C_s > 0.044 S_{DS} I_e \quad \text{or} \quad 0.01, \quad \text{whichever is larger} \qquad (8.13)$$

or

$$C_s > \frac{0.5 S_1}{R/I_e}, \quad \text{for} \quad S_1 \ge 0.6g \qquad (8.14)$$

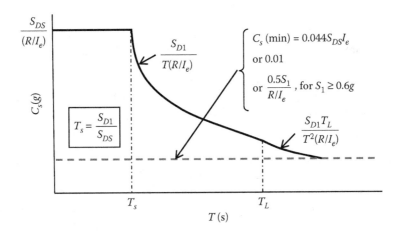

FIGURE 8.6 Inelastic design response spectrum.

where:

S_{D1} is the design spectral response acceleration at a period of 1 s, computed using the second of Equations 8.8 (ASCE-7, Section 11.4.4)

T is the fundamental period of the structure computed in step 2

T_L is the long-period transition period (ASCE-7, Section 11.4.5)

S_1 is the MCE spectral response acceleration at a period of 1 s discussed in Section 8.4.1 (ASCE-7, Section 11.4.4)

The two limits in Equation 8.13 are intended to represent the minimum base shear force levels to safeguard against the collapse of long-period structures (uncertainty related to P-Δ effects), while Equation 8.14 accounts for the effects of near-fault directivity.

4. Compute the equivalent lateral forces (ASCE-7, Section 12.8.3). The lateral seismic force, F_x, induced at any level x as illustrated in Figure 8.7, can be determined using the following relationship:

$$F_x = C_{vx}V \tag{8.15}$$

where:

V is the base shear computed in step 3

C_{vx} is the vertical distribution factor

$$C_{vx} = \frac{w_x \cdot h_x^k}{\sum_{i=1}^{n} w_i \cdot h_i^k} \tag{8.16}$$

where:

w_x and w_i are the weights at level x or i, depicted in Figure 8.7

h_x and h_i are the heights from the base to the level x or i, depicted in Figure 8.7

n is the number of stories

k depends on the structural period and accounts for higher mode effects

$k = 1$ for $T \leq 0.5$ s

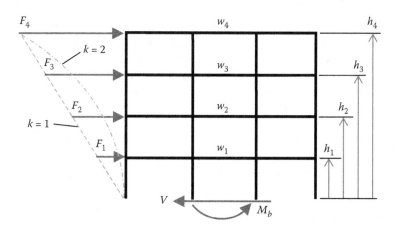

FIGURE 8.7 Equivalent lateral forces and overturning moment.

$k = 2$ for 0.5 s $< T < 2.5$ s, or can be determined by linear interpolation between 1 and 2

$k = 2$ for $T \geq 2.5$ s

These formulas are approximately the same as Equation 6.37 for a generalized SDOF shear building, where we assumed that the lateral response is dominated by the fundamental mode. The shape vector, ψ_i, in this case is assumed to vary from linear ($k = 1$) to parabolic ($k = 2$), as shown in Figure 8.7. The parabolic mode is intended to account for the response of flexible structures.

5. Compute overturning moment (ASCE-7, Section 12.8.5). As discussed in Section 6.1.3, with the equivalent static forces, we can determine the overturning moment by applying static equilibrium to the free-body diagram depicted in Figure 8.7, that is,

$$M_b = \sum_{i=1}^{n} h_i F_i \tag{8.17}$$

As shown in Figure 6.3, we can also obtain the internal story shear force, $V_{\max i}$, and the internal story moment, $M_{\max i}$, at an arbitrary level i by applying static equilibrium to the right-hand side free-body diagram as shown by Equations 6.38 and 6.39, respectively, and are repeated as follows for convenience:

$$V_{\max i} = \sum_{j=i}^{N} f_{jo}$$

$$M_{\max i} = \sum_{j=i}^{N} (h_j - h_i) f_{jo}$$

Again, in the derivation of Equations 8.17 and 6.39, we assumed that the floor weights are directly in the center of the building frame, thus making no contribution to the moment equilibrium equation. However, for buildings with off-center weights, a full moment static equilibrium analysis is conducted to account for the contribution of these off-center forces.

6. Determine story drifts (ASCE-7, Section 12.8.6). With the lateral forces, we can perform a structural analysis to determine lateral displacements. The relative displacement of each story at the center of mass is defined as the drift. For cases where centers of mass do not align and for those cases with irregularities, restrictions are specified in ASCE-7. Note that the loads used to calculate drifts are at strength level (not service), as indicated in ASCE-7, Figure 12.8-2. The drift is computed using the following inelastic displacement at level x:

$$\delta_x = \frac{C_d \delta_{xe}}{I_e} \tag{8.18}$$

where:
C_d is the deflection amplification factor discussed in Section 8.3.2 and given in ASCE-7, Table 12.2-1

I_e is the importance factor introduced in Section 8.4.2 and used in step 3
δ_{xe} is the deflection at level x determined from an elastic analysis

The drift is then computed as

$$\Delta_x = \delta_x - \delta_{x-1} \tag{8.19}$$

where:
  δ_x is the deflection at the top of level x determined using Equation 8.18
  δ_{x-1} is the deflection at the bottom of level x determined using Equation 8.18

To ensure the stability of the overall system (i.e., avoid P-Δ effects) and to minimize damage to nonstructural elements, these drifts must be kept within allowable limits per ASCE-7, Section 12.12.1, which are listed in Table 12.12-1 and depend on the structure type and occupancy category.

When P-Δ effects are significant, provisions in ASCE-7, Section 12.9.6, should be followed to ensure stability. For cases where torsional effects are significant, ASCE-7, Section 12.8.4.3, requires the application of torsional amplification factors in the analysis.

EXAMPLE 8.5

Given the following three-story office building frame located on the campus of University of the Pacific in Stockton, California, compute the approximate structural period T_a, the design base shear V, the design earthquake loads acting on each floor F_x, and the overturning moment at the base. The building in plan view is square; the structural system being proposed is a special steel concentrically braced frame (SSCBF) as shown in Figure E8.5 with an estimated weight on each of the first two floors of 100 and 80 kips on the roof. The geotechnical engineer has classified the site (soil) class as D.

SOLUTION

1. Determine the seismic weight. The weight of the roof and two floors is given; the total weight of the system is the sum of these weights,

$$W = 100\,\text{kips} + 100\,\text{kips} + 80\,\text{kips} = 280\,\text{kips}$$

2. Determine the period. No information about structural stiffness is given, so we use the approximate period given by Equation 8.9,

$$T_a = C_t h_n^x$$

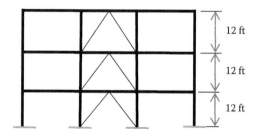

FIGURE E8.5 Elevation view of the special steel concentrically braced frame.

where:

$h_n = 36$ ft and is the building height

for an SSCBF, which corresponds to all other structural systems in Table 3.2,

$C_t = 0.02$

$x = 0.75$

Thus,

$$T_a = C_t h_n^x = 0.02(36 \text{ ft})^{0.75} = 0.29 \text{ s}$$

Note that this value is very close to the result of Equation 3.19, $T_a = 0.1N = 0.1(3) = 0.3$ s. This relationship can be used to check the validity of the calculated values of the period.

3. Calculate the base shear, $V = C_s W$ per ASCE-7, Section 12.8. The base shear is obtained using Equation 8.10, with Equations 8.11 and 8.12, or the response spectrum shown in Figure 8.6. Using the spectral values obtained in Example 8.4, draw the response spectrum for this case as shown in Figure E8.6.

With $T_a = 0.29$ s, we can enter this spectrum. Since $T_o < T_a \leq T_s$, $S_a = S_{DS} = 0.65$ and

$$C_s = \frac{S_{DS}}{R/I_e}$$

As noted in the problem statement, the building will be used for office space, which corresponds to an occupancy category of II; thus, the importance factor obtained from Table 8.2 is

$$I_e = 1.0$$

Also, we can determine the SDC using Table 8.1 (or the USGS application discussed in Example 8.4) to ensure that the selected seismic load-resisting system is permitted, that

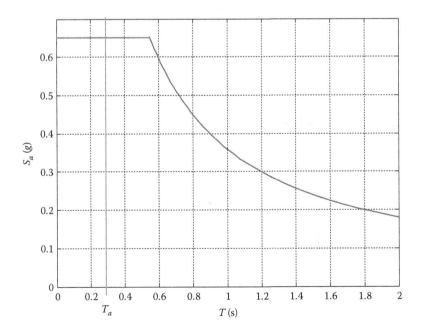

FIGURE E8.6 Design acceleration response spectrum.

is, for Occupancy Category II, and $S_{DS} = 0.65 > 0.5$ and $S_{D1} = 0.36 > 0.2$, the governing SDC is D, as shown in the following table.

Based on S_{DS} (Table 11.6-1)	Based on S_{D1} (Table 11.6-2)	I or II or III	IV
$S_{DS} < 0.167$	$S_{D1} < 0.067$	A	A
$0.167 \leq S_{DS} < 0.33$	$0.067 \leq S_{D1} < 0.133$	B	C
$0.33 \leq S_{DS} < 0.50$	$0.133 \leq S_{D1} < 0.20$	C	D
$S_{DS} \geq 0.50$	$S_{D1} \geq 0.20$	D	D
	$S_1 \geq 0.75$	E	F

Thus, for SSCBF (ASCE-7, Table 12.2-1, case 2 in part B), the height limit is 160 ft, which is larger than the proposed 36 ft building height. The seismic response coefficient is also obtained from ASCE-7, Table 12.2-1,

$$R = 6.0$$

The base shear is calculated as follows:

$$V = C_s W = \frac{S_{DS}}{R/I_e} W = \frac{0.65}{6.0/1.0}(280 \text{ kips}) = 30.7 \text{ kips}$$

4. Determine lateral forces at each level of the structure, $F_x = C_{vx}V$ per ASCE-7, Section 12.8.3. The forces are obtained using Equations 8.15 and 8.16 for each level, x:

$$F_x = C_{vx}V$$

where:
  $V = 30.7$ kips, computed in step 3
  C_{vx} is the vertical distribution factor with $k = 1$ since $T_a = 0.29 \text{ s} \leq 0.5 \text{ s}$

$$C_{vx} = \frac{w_x \cdot h_x}{\sum_{i=1}^{3} w_i \cdot h_i} = \frac{w_x \cdot h_x}{100 \text{ kips}(12 \text{ ft}) + 100 \text{ kips}(24 \text{ ft}) + 80 \text{ kips}(36 \text{ ft})} = \frac{w_x \cdot h_x}{6480 \, k \cdot \text{ft}}$$

So the lateral forces are:

$$F_1 = C_{v1}V = \frac{w_1 \cdot h_1}{6480 \, k \cdot \text{ft}}(30.7 \text{ kips}) = \frac{100 \text{ kips}(12 \text{ ft})}{6480 \, k \cdot \text{ft}}(30.7 \text{ kips}) = 5.68 \text{ kips}$$

$$F_2 = C_{v2}V = \frac{w_2 \cdot h_2}{6480 \, k \cdot \text{ft}}(30.7 \text{ kips}) = \frac{100 \text{ kips}(24 \text{ ft})}{6480 \, k \cdot \text{ft}}(30.7 \text{ kips}) = 11.36 \text{ kips}$$

$$F_3 = C_{v3}V = \frac{w_3 \cdot h_3}{6480 \, k \cdot \text{ft}}(30.7 \text{ kips}) = \frac{80 \text{ kips}(36 \text{ ft})}{6480 \, k \cdot \text{ft}}(30.7 \text{ kips}) = 13.63 \text{ kips}$$

5. Determine the overturning moment using static equilibrium. With the lateral forces computed in the last step, we can determine the overturning moment at the base using Equation 8.17,

$$M_b = \sum_{i=1}^{3} h_i F_i = 5.68 \text{ kips}(12 \text{ ft}) + 11.36 \text{ kips}(24 \text{ ft}) + 13.63 \text{ kips}(36 \text{ ft}) = 831 \text{ kip} \cdot \text{ft}$$

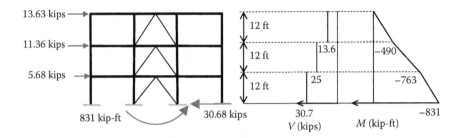

FIGURE E8.7 Summary of analysis results using the equivalent lateral force method.

We can also obtain the internal story shear forces and moments at each level using Equations 6.38 and 6.39, respectively. Alternatively, we can draw shear force and bending moment diagrams by treating the building as a cantilever beam in order to obtain the internal story shear forces and moments. Figure E8.7 summarizes the results of the analysis.

EXAMPLE 8.6

Check the allowable story drifts for the three-story office building frame given in Example 8.5. A structural analysis of the building frame was conducted with lateral forces determined in Example 8.5 in order to determine the elastic displacements at each level ($\delta_{1e} = 0.65$ in, $\delta_{2e} = 1.3$ in, and $\delta_{3e} = 1.6$ in) and illustrated in Figure E8.8.

SOLUTION

1. Determine inelastic displacements per ASCE-7, Section 12.8.6. Compute inelastic displacements at each level x using Equation 8.18,

$$\delta_x = \frac{C_d \delta_{xe}}{I_e}$$

where:

$C_d = 5.0$ for steel special concentrically braced frames given in ASCE-7, Table 12.2-1
$I_e = 1$ for the office building
δ_{xe} is the deflection at level x given in Figure E8.8.

So the inelastic displacements are:

$$\delta_1 = C_d \delta_{1e}/I_e = 5(0.65 \text{ in})/1 = 3.25 \text{ in}$$

$$\delta_2 = C_d \delta_{2e}/I_e = 5(1.3 \text{ in})/1 = 6.5 \text{ in}$$

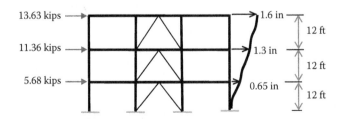

FIGURE E8.8 Lateral forces and elastic displacements at each level of the building frame.

$$\delta_3 = C_d\delta_{3e}/I_e = 5(1.6 \text{ in})/1 = 8.0 \text{ in}$$

2. Determine story drifts. Compute drifts at each level using Equation 8.19,

$$\Delta_1 = \delta_1 - \delta_{1-1} = \delta_1 - \delta_0 = 3.25 \text{ in} - 0 = 3.25 \text{ in}$$

$$\Delta_2 = \delta_2 - \delta_{2-1} = \delta_2 - \delta_1 = 6.5 \text{ in} - 3.25 \text{ in} = 3.25 \text{ in}$$

$$\Delta_3 = \delta_3 - \delta_{3-1} = \delta_3 - \delta_2 = 8 \text{ in} - 6.5 \text{ in} = 1.5 \text{ in}$$

3. Check the allowable story drifts per ASCE-7, Section 12.12.1. For buildings in Occupancy Category II, ASCE-7, Table 12.12-1, gives an allowable drift (for four stories or less structures with interior walls, partitions, etc. designed to accommodate story drifts) of

$$\Delta_a = 0.025\, h_{sx}$$

where:
  h_{sx} is the story height for level x. In this case, each level is 12 ft high; thus,

$$\Delta_a = 0.025\, h_{sx} = 0.025(12 \text{ ft})(12 \text{ in/ft}) = 3.6 \text{ in}$$

None of the story drifts exceed the allowable drift.

8.6   MODAL RESPONSE SPECTRUM ANALYSIS

As discussed in Section 8.4, earthquake loads can be obtained using one of two structural dynamics-based methods, the MRS (discussed in this section) or response time-history procedure (briefly covered in the next section) analyses. ASCE-7 MRS analysis is roughly based on the MDOF system analysis for shear buildings presented in Chapter 7. After determining the modal properties for each mode (natural period, shape, modal participation factor, and effective modal weight), the remainder of the process is similar to the ELFP method in that it involves determining the base shear, equivalent static story forces, story drifts, and overturning moments for each participating mode. The number of participating modes in each horizontal direction should be enough to obtain the modal weight participation of at least 90% of the actual weight.

The ASCE-7 MRS analysis steps are as follows and are similar to the summary presented in Section 7.2 (corresponding ASCE-7 sections are given in parentheses). Note that many of the steps required for the ELFP method are the same as the ones that follow, and many of the additional steps are usually performed by a computer.

1. Determine the mass matrix $[m]$ from the floor and roof seismic weight using details presented in Section 3.1.2 (ASCE-7, Section 12.7.2).
2. Determine the stiffness matrix $[k]$ from the properties of the structural system using details presented in Section 7.1.
3. With $[m]$ and $[k]$, solve the associated eigenvalue problem to determine modal properties for each mode using details presented in Section 7.1.1:
   a. Natural frequencies ω_{nj} and periods T_{nj} using the eigenvalues ω_j^2.
   b. Mode shapes ϕ_{ij} using normalized eigenvectors.

c. Modal participation factors Γ_j (which are the contribution of each mode shape to the total response) using Equation 7.50, repeated here for convenience,

$$\Gamma_j = \frac{L_j}{M_j} = \frac{\sum_{i=1}^{n} m_i \phi_{ij}}{\sum_{i=1}^{n} m_i \phi_{ij}^2} \quad \text{or} \quad \{\Gamma\} = \frac{[\Phi]^T [m]\{1\}}{[\Phi]^T [m][\Phi]}$$

Where we have assumed that $m_{ii} = m_i$ since the mass matrix is diagonal.

d. Effective modal weight using Equation 7.66, repeated here for convenience,

$$W_j^e = \frac{\left(\sum_{i=1}^{n} w_i \phi_{ij} \right)^2}{\sum_{i=1}^{n} w_i \phi_{ij}^2} \quad \text{or} \quad W_j^e = \Gamma_j \sum_{i=1}^{n} w_i \phi_{ij} \quad \text{or} \quad W_j^e = \Gamma_j L_j g$$

where:

$w_i = m_i \cdot g$, which is the fraction of the building seismic weight at level i

n is the number of stories

4. Determine the number of modes to be used in the analysis (ASCE-7, Section 12.9.1). Use a sufficient number of modes ϕ_{ij} in each direction to account for at least 90% of the actual weight W.

5. Develop the design response spectrum and determine the spectral accelerations for participating modes.
   a. Develop an elastic response spectrum using details presented in Section 8.4.1, Figure 8.5 (ASCE-7, Section 11.4.5).
   b. Determine spectral accelerations for the mth mode, S_{am}, using the elastic response spectrum and the corresponding periods computed in step 3 for each contributing mode found in step 4.
   c. Divide S_{am} by the quantity R/I_e to obtain the modal seismic response coefficient,

$$C_{sm} = \frac{S_{am}}{R/I_e} \qquad (8.20)$$

where:

R is the response modification factor in ASCE-7, Table 12.2-1

I_e is the importance factor in Table 8.2 (ASCE-7, Section 11.5.1)

6. Determine the system base shear, V_t.
   a. Compute seismic modal base shear V_m for each of the participating modes; that is, the fraction of the total base shear contributed by the mth mode of vibration is

$$V_m = C_{sm} W_m = C_{sm} \Gamma_m \sum_{i=1}^{n} w_i \phi_{im} \qquad (8.21)$$

where:

C_{sm} is given by Equation 8.20 and is usually expressed as a fraction of g

W_m is the effective modal weight for the participating modes computed in step 3d

w_i is the weight located at level i

ϕ_{im} is the displacement at the ith level for the mth mode

b. Statistically combine the modal base shears using SRSS (Equation 7.68) or CQC (Equations 7.69 and 7.70) to obtain the system base shear V_t (Section 12.9.3).

c. Ensure that V_t is larger than a minimum value based on Section 8.5 ELFP analysis, V (ASCE-7, Section 12.9.4.1). If V_t is less than 85% of V (with $T = C_u T_a$ when $T > C_u T_a$), then lateral force, story shears, and story moments determined using the modal analysis must be multiplied by $0.85 \, V/V_t$. C_u and T_a are discussed in Section 8.5.

7. Determine the lateral forces at each level x (ASCE-7, Section 12.8.3). The fraction of the total lateral force contributed by the mth mode of vibration at each level x, F_{xm}, is

$$F_{xm} = C_{vxm} V_m \qquad (8.22)$$

where:

V_m is the seismic modal base shear computed in step 6a

C_{vxm} is the vertical distribution factor

$$C_{vxm} = \frac{w_x \phi_{xm}}{\sum_{i=1}^{n} w_i \phi_{im}} \qquad (8.23)$$

where:

w_x and w_i are the weights at levels x and i, depicted in Figure 8.7

ϕ_{im} and $\phi_{xm} = i$th and xth level displacements for the mth mode

We can then statistically combine these modal lateral forces using SRSS (Equation 7.68) or CQC (Equations 7.69 and 7.70) to obtain the design lateral forces at each level F_x (ASCE-7, Section 12.9.3).

8. Compute overturning moment. With the modal lateral forces obtained in the last step F_{xm}, we can determine the modal overturning moments by applying static equilibrium,

$$M_{bm} = \sum_{i=1}^{n} h_i F_{im} \qquad (8.24)$$

We can then statistically combine these modal overturning moments using SRSS (Equation 7.68) or CQC (Equations 7.69 and 7.70) to obtain the design overturning moment at the base M_b (ASCE-7, Section 12.9.3).

9. Determine story drifts. Story drifts are the relative displacements of each story, computed by taking the difference between displacements of two consecutive floors. Note that the modal lateral loads used to calculate drifts are at the strength level (not service), as indicated in ASCE-7, Figure 12.8-2.

a. The fraction of modal inelastic displacement contributed by the mth mode of vibration at level x is

$$\delta_{xm} = \frac{C_d \delta_{xem}}{I_e} \qquad (8.25)$$

where:

C_d is the deflection amplification factor discussed in Section 8.3.2 and given in ASCE-7, Table 12.2-1

I_e is the importance factor introduced in Section 8.4.2

δ_{xem} is the deflection at level x in the mth mode, and can be determined using the following relationship:

$$\delta_{xem} = \left(\frac{g}{4\pi^2} \right) \frac{T_m^2 F_{xm}}{w_x} \tag{8.26}$$

where:

g is the acceleration due to gravity

T_m is the period of the mth mode obtained following step 3a

w_x is the weight at level x

F_{xm} is the modal lateral forces determined in step 7

b. Statistically combine the modal inelastic displacements using SRSS (Equation 7.68) or CQC (Equations 7.69 and 7.70) to obtain the design displacements.

c. The drift is then computed as

$$\Delta_{xm} = \delta_{xm} - \delta_{xm-1} \tag{8.27}$$

where:

δ_{xm} is the deflection at the top of level x determined using Equation 8.25

δ_{xm-1} is the deflection at the bottom of level x determined using Equation 8.25

d. Statistically combine the modal drifts using SRSS (Equation 7.68) or CQC (Equations 7.69 and 7.70) to obtain the design drifts.

EXAMPLE 8.7

A four-story office building frame, with equal story heights of 12 ft, is located at a site with site soil class D, $S_{D1} = 0.75$ and $S_{DS} = 1.33$. The structural system is a special reinforced concrete moment frame with the following stiffness matrix [k] and story weights of $w_1 = w_2 = w_3 = 789$ kips and $w_4 = 645$ kips (Villaverde 2009). Use the MRS analysis to compute the design earthquake loads acting on each floor, the story drifts, and the overturning moment:

$$[k] = \begin{bmatrix} 5611 & -2916 & 458 & -45 \\ -2916 & 3863 & -1885 & 269 \\ 458 & -1885 & 2893 & -1398 \\ -45 & 269 & -1398 & 1168 \end{bmatrix} \text{kips/in} \quad \text{and} \quad [w] = \begin{bmatrix} 789 & 0 & 0 & 0 \\ 0 & 789 & 0 & 0 \\ 0 & 0 & 789 & 0 \\ 0 & 0 & 0 & 645 \end{bmatrix} \text{kips}$$

SOLUTION

1. Determine the mass matrix [m]. Use the floor and roof seismic weights given in the problem statement (the matrix can be obtained using D'Alembert's principle as shown in Chapter 7, Example 8.4). The first floor is listed in the first row, while the roof is listed along the last row (some books will reverse this order).

$$[m] = [w]/g = \begin{bmatrix} 2.04 & 0 & 0 & 0 \\ 0 & 2.04 & 0 & 0 \\ 0 & 0 & 2.04 & 0 \\ 0 & 0 & 0 & 1.67 \end{bmatrix} \text{kips} \cdot \text{s}^2/\text{in}$$

where:

$g = 386.4 \text{ in/s}^2$

2. Determine the stiffness matrix $[k]$. Use the stiffness matrix given in the problem statement (this matrix can be obtained using techniques from matrix structural analysis).
3. With $[m]$ and $[k]$, solve the associated eigenvalue problem to determine modal properties for each mode using MATLAB®:
   a. Natural frequencies ω_{nj} and periods T_{nj} using the eigenvalues ω_j^2.

$$\{\omega_n\} = \begin{Bmatrix} 8.8 \\ 26.1 \\ 44.1 \\ 63.7 \end{Bmatrix} \text{rad/s} \quad \text{and} \quad \{T_n\} = \left\{ \frac{2\pi}{\omega_n} \right\} = \begin{Bmatrix} 0.713 \\ 0.241 \\ 0.143 \\ 0.099 \end{Bmatrix} \text{s}$$

   b. Mode shapes ϕ_{ij} using the normalized eigenvectors $[\Phi]$.

$$[\Phi] = \begin{bmatrix} 0.2191 & -0.6832 & 1.0193 & -6.4994 \\ 0.5257 & -1.0288 & 0.3319 & 5.5143 \\ 0.8369 & -0.1517 & -1.4521 & -2.7391 \\ 1.0 & 1.0 & 1.0 & 1.0 \end{bmatrix}$$

   c. Modal participation factors $\{\Gamma\}$ using Equation 7.50, $\{\Gamma\} = [\Phi]^T[m]\{1\}/[\Phi]^T[m][\Phi]$. First, determine the modal mass matrix, $[M] = [\Phi]^T[m][\Phi]$,

$$[M] = \begin{bmatrix} 0.2191 & 0.5257 & 0.8369 & 1.0 \\ -0.6832 & -1.0288 & -0.1517 & 1.0 \\ 1.0193 & 0.3319 & -1.4521 & 1.0 \\ -6.4994 & 5.5143 & -2.7391 & 1.0 \end{bmatrix} \begin{bmatrix} 2.04 & 0 & 0 & 0 \\ 0 & 2.04 & 0 & 0 \\ 0 & 0 & 2.04 & 0 \\ 0 & 0 & 0 & 1.67 \end{bmatrix}$$

$$\times \begin{bmatrix} 0.2191 & -0.6832 & 1.0193 & -6.4994 \\ 0.5257 & -1.0288 & 0.3319 & 5.5143 \\ 0.8369 & -0.1517 & -1.4521 & -2.7391 \\ 1.0 & 1.0 & 1.0 & 1.0 \end{bmatrix} \text{kips} \cdot \text{s}^2/\text{in}$$

$$[M] = \begin{bmatrix} 3.76 & 0 & 0 & 0 \\ 0 & 4.83 & 0 & 0 \\ 0 & 0 & 8.32 & 0 \\ 0 & 0 & 0 & 165.35 \end{bmatrix} \text{kips} \cdot \text{s}^2/\text{in}$$

And the modal force, $[L] = [\Phi]^T[m]\{1\}$,

$$[L] = \begin{bmatrix} 0.2191 & 0.5257 & 0.8369 & 1.0 \\ -0.6832 & -1.0288 & -0.1517 & 1.0 \\ 1.0193 & 0.3319 & -1.4521 & 1.0 \\ -6.4994 & 5.5143 & -2.7391 & 1.0 \end{bmatrix} \begin{bmatrix} 2.04 & 0 & 0 & 0 \\ 0 & 2.04 & 0 & 0 \\ 0 & 0 & 2.04 & 0 \\ 0 & 0 & 0 & 1.67 \end{bmatrix} \begin{Bmatrix} 1 \\ 1 \\ 1 \\ 1 \end{Bmatrix} \text{kips} \cdot \text{s}^2/\text{in}$$

$$[L] = \begin{Bmatrix} 4.90 \\ 2.14 \\ 1.46 \\ -5.94 \end{Bmatrix} \text{kips} \cdot \text{s}^2/\text{in}$$

Next determine the participation factors,

$$\{\Gamma\} = [M]^{-1}\{L\} = \begin{bmatrix} 1/3.76 & 0 & 0 & 0 \\ 0 & 1/4.83 & 0 & 0 \\ 0 & 0 & 1/8.32 & 0 \\ 0 & 0 & 0 & 1/165.35 \end{bmatrix} \text{kips} \cdot \text{s}^2/\text{in} \begin{Bmatrix} 4.90 \\ -2.14 \\ 1.46 \\ -5.94 \end{Bmatrix} \text{kips} \cdot \text{s}^2/\text{in}$$

$$\{\Gamma\} = \begin{Bmatrix} 1.3023 \\ -0.4422 \\ 0.1758 \\ -0.0359 \end{Bmatrix}$$

d. Calculate the effective modal weight using Equation 7.66, $W_j^e = \Gamma_j \sum_{i=1}^{n} w_i \phi_{ij}$.

$$W_1^e = \Gamma_1 \sum_{i=1}^{4} w_i \phi_{i1} = 1.3023(789(0.219 + 0.526 + 0.837) + 645(1)) = 2466 \text{ kips}$$

Similarly,

$$W_2^e = 365 \text{ kips}$$

$$W_3^e = 99 \text{ kips}$$

$$W_4^e = 82 \text{ kips}$$

4. Determine the number of modes to be used in the analysis. Use enough modes ϕ_{ij} in each direction to capture at least 90% of the actual weight W. The total weight of the system is $W = 4(789 \text{ kips}) + 645 \text{ kips} = 3012 \text{ kips}$. The percentages of the total weight per mode are as follows:
   Mode 1, $W_1^e/W \times 100 = 2466 \text{ kips}/3012 \text{ kips} = 81.8\%$
   Mode 2, $W_2^e/W \times 100 = 365 \text{ kips}/3012 \text{ kips} = 12.1\%$
   Mode 3, $W_3^e/W \times 100 = 99 \text{ kips}/3012 \text{ kips} = 3.3\%$
   Mode 4, $W_4^e/W \times 100 = 82 \text{ kips}/3012 \text{ kips} = 2.7\%$
   Therefore, we only need to include the first two modes, which capture 93.9% of W.
5. Develop the design response spectrum and determine spectral accelerations for partici-pating modes.
   a. Develop an elastic response spectrum using details presented in Section 8.4.1, Figure 8.5.

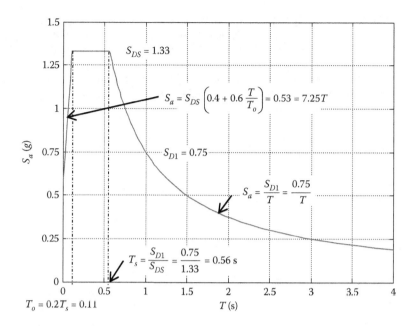

FIGURE E8.9 Design acceleration response spectrum.

b. Determine spectral accelerations using Figure E8.9 and the periods computed in step 3 for each contributing mode found in step 4. Enter the graph with T_1 and T_2 (since only the first two modes are required) to acquire spectral accelerations for the two participating modes:

$T_1 = 0.713$ s $> T_s = 0.56$ s; therefore, $S_{a1} = S_{D1}/T_1 = 0.75/0.713 = 1.05$
$T_2 = 0.241$ s $< T_s = 0.56$ s; therefore, $S_{a2} = 1.33$

c. Divide S_{am} by R/I_e to obtain the design modal seismic response coefficients, C_{sm}, $I_e = 1.0$ from Table 8.2, since an office building is classified with an Occupancy Category II. $R = 8$, for a special moment reinforced concrete frame (case C5 in ASCE-7, Table 12.2-1).

Thus,

$$C_{s1} = \frac{S_{a1}}{R/I_e} = \frac{1.05}{8/1} = 0.13$$

$$C_{s2} = \frac{S_{a2}}{R/I_e} = \frac{1.33}{8/1} = 0.17$$

6. Determine the system base shear, V_b.
    a. Compute the seismic modal base shear for each of the participating modes with C_{sm} from step 5c and W_m from step 3d:

$$V_1 = C_{s1}W_1 = 0.13(2465 \text{ kips}) = 320 \text{ kips}$$

$$V_2 = C_{s2}W_2 = 0.17(365 \text{ kips}) = 62 \text{ kips}$$

    b. Combine V_1 and V_2 using SRSS (Equation 7.68) to obtain the base shear $V_t = \sqrt{\Sigma V_i^2}$.

$$V_t = \sqrt{(320 \text{ kips})^2 + (62 \text{ kips})^2} = 326 \text{ kips}$$

c. Verify that V_t is larger than a minimum base shear value, V, obtained with an ELFP analysis described in Section 8.5. If V_t is less than 85% of V (with $T = C_uT_a$ when $T > C_uT_a$), then lateral force, story shears, and story moments determined using the modal analysis should be multiplied by 0.85 V/V_t.

First, calculate the base shear V using the ELFP method, with $T = C_uT_a$:

$C_u = 1.4$ since $S_{D1} > 0.4$ (ASCE-7, Table 12.8-1)

$T_a = C_th_n^x$ as discussed in Section 8.5

where:

$h_n = 48$ ft and is the building height

for a concrete moment-resisting frame in Table 3.2 (ASCE-7, Table 12.8-2)

$$C_t = 0.016$$

$$x = 0.9$$

So,

$$T_a = C_th_n^x = 0.016(48 \text{ ft})^{0.9} = 0.53 \text{ s}$$

Thus,

$$C_uT_a = 1.4(0.53 \text{ s}) = 0.73 \text{ s} > T_s = 0.56 \text{ s}; \quad \text{therefore, } S_a = S_D/T = 0.75/0.73 = 0.13$$

and

$$V = C_sW = 0.13(3012 \text{ kips}) = 387 \text{ kips}$$

Furthermore,

$$0.85 \ V = 0.85(387 \text{ kips}) = 329 \text{ kips} > V_t = 326 \text{ kips}$$

Therefore, we must multiply V_t by 0.85 V/V_t to obtain the actual design base shear:

$$V_b = (V_t)(0.85 \ V/V_t) = 0.85 \ V = 329 \text{ kips}$$

7. Determine the lateral forces, F_x.
   a. Use Equations 8.22 through 8.24 to determine the modal lateral force F_{xm}: Mode 1 forces at each level x,

   $$F_{x1} = C_{vx1}V_1$$

where:

$V_1 = 0.85 \ V/V_t$ (320 kips) = 1.01(320 kips) = 323 kips, computed in step 6a

$$C_{vx1} = \frac{w_x\phi_{x1}}{\sum_{i=1}^{n} w_i\phi_{i1}} = \frac{w_x\phi_{x1}}{789 \text{ kips}(0.219 + 0.526 + 0.837) + 645 \text{ kips}} = \frac{w_x\phi_{x1}}{1890 \ k \cdot \text{ft}}$$

So the lateral forces are:

$$F_{11} = \frac{w_1\phi_{11}}{1890 \ k \cdot \text{ft}}(323 \text{ kips}) = \frac{789 \text{ kips}(0.219)}{1890 \ k \cdot \text{ft}}(323 \text{ kips}) = 30 \text{ kips}$$

$$F_{21} = \frac{w_2\phi_{21}}{1890\ k\cdot ft}(323\ \text{kips}) = \frac{789\ \text{kips}(0.526)}{1890\ k\cdot ft}(323\ \text{kips}) = 71\ \text{kips}$$

$$F_{31} = \frac{w_3\phi_{31}}{1890\ k\cdot ft}(323\ \text{kips}) = \frac{789\ \text{kips}(0.837)}{1890\ k\cdot ft}(323\ \text{kips}) = 113\ \text{kips}$$

$$F_{41} = \frac{w_4\phi_{41}}{1890\ k\cdot ft}(323\ \text{kips}) = \frac{641\ \text{kips}(1)}{1890\ k\cdot ft}(323\ \text{kips}) = 110\ \text{kips}$$

Mode 2 forces at each level x are similarly computed:

$$F_{x2} = C_{vx2}V_2$$

where:

$V_2 = 0.85\ V/V_t\ (62\ \text{kips}) = 1.01(62\ \text{kips}) = 62.6\ \text{kips}$, computed in step 6a

$$C_{vx2} = \frac{w_x\phi_{x2}}{\sum_{i=1}^{n} w_i\phi_{i2}} = \frac{w_x\phi_{x2}}{789\ \text{kips}(-0.683-1.029-0.152)+645\ \text{kips}} = \frac{w_x\phi_{x2}}{-826k\cdot ft}$$

So the lateral forces are:

$$F_{12} = \frac{w_1\phi_{12}}{-826\ k\cdot ft}(62.6\ \text{kips}) = \frac{789\ \text{kips}(-0.152)}{-826\ k\cdot ft}(62.6\ \text{kips}) = 41\ \text{kips}$$

$$F_{22} = \frac{w_2\phi_{22}}{-826\ k\cdot ft}(62.6\ \text{kips}) = \frac{789\ \text{kips}(-1.029)}{-826\ k\cdot ft}(62.6\ \text{kips}) = 61\ \text{kips}$$

$$F_{32} = \frac{w_3\phi_{32}}{-826\ k\cdot ft}(62.6\ \text{kips}) = \frac{789\ \text{kips}(-0.683)}{-826\ k\cdot ft}(62.6\ \text{kips}) = 9.1\ \text{kips}$$

$$F_{42} = \frac{w_4\phi_{42}}{-826k\cdot ft}(62.6\ \text{kips}) = \frac{641\ \text{kips}(1)}{-826\ k\cdot ft}(62.6\ \text{kips}) = -48\ \text{kips}$$

b.   Combine lateral forces for mode 1, F_{x1}, and mode 2, F_{x2}, using SRSS (Equation 7.68) to obtain the lateral forces, $F_x = \sqrt{\Sigma F_{xi}^2}$.

$$F_1 = \sqrt{(30\ \text{kips})^2 + (41\ \text{kips})^2} = 51\ \text{kips}$$

$$F_2 = \sqrt{(71\ \text{kips})^2 + (61\ \text{kips})^2} = 94\ \text{kips}$$

$$F_3 = \sqrt{(113\ \text{kips})^2 + (9.1\ \text{kips})^2} = 113\ \text{kips}$$

$$F_4 = \sqrt{(110\ \text{kips})^2 + (-48\ \text{kips})^2} = 120\ \text{kips}$$

8. Determine the overturning moment.
   a. With the modal lateral forces obtained in the last step, F_{xm}, we can determine the modal overturning moments by applying static equilibrium for each mode: The mode 1 overturning moment at the base is

$$M_{b1} = \sum_{i=1}^{4} h_i F_{i1} = 30(12) + 71(24) + 113(36) + 110(48) = 11,412 \, k \cdot ft$$

Mode 2 overturning moment at the base:

$$M_{b2} = \sum_{i=1}^{4} h_i F_{i2} = 41(12) + 61(24) + 9.1(36) - 48(48) = -20 \, k \cdot ft$$

   b. Combine M_{b1} and M_{b2} using SRSS (Equation 7.68) to obtain the system overturning moment at the base $M_b = \sqrt{\Sigma M_{bi}^2}$.

$$M_b = \sqrt{(11,412 \, k \cdot ft)^2 + (-20 \, k \cdot ft)^2} = 11,412 \, k \cdot ft$$

Note the factor 0.85 V/V_t was included in the forces.
9. Determine story drifts.
   a. The modal inelastic displacement, δ_{xm}, at level x is

$$\delta_{xm} = \frac{C_d \delta_{xem}}{I_e}$$

where:

$C_d = 5.5$ for special moment reinforced concrete frames given in ASCE-7, Table 12.2-1
$I_e = 1$ for an office building

$$\delta_{xem} = \left(\frac{g}{4\pi^2}\right) \frac{T_m^2 F_{xm}}{w_x}$$

where:

T_m is the period of the mth mode obtained in step 3a
w_x is the weight at level x from step 1
F_{xm} is the modal lateral forces obtained in step 7a without the factor 0.85(V/V_t)

Mode 1 inelastic displacement is calculated as follows:

$$\delta_{x1} = \frac{C_d}{I_e} \delta_{xe1} = \frac{5.5}{1} \left(\frac{386.4 \, in/s^2}{4\pi^2}\right) \frac{(0.713 \, s)^2 F_{x1}}{w_x} = 27.37 \, in \frac{F_{x1}}{w_x}$$

The inelastic displacements at each level x are then determined as follows:

$$\delta_{11} = 27.37 \, in \frac{F_{11}}{w_1} = 27.37 \, in \frac{29.3 \, kips}{789 \, kips} = 1.02 \, in$$

$$\delta_{21} = 27.37 \text{ in} \frac{F_{21}}{w_2} = 27.37 \text{ in} \frac{70.3 \text{ kips}}{789 \text{ kips}} = 2.44 \text{ in}$$

$$\delta_{31} = 27.37 \text{ in} \frac{F_{31}}{w_3} = 27.37 \text{ in} \frac{111.8 \text{ kips}}{789 \text{ kips}} = 3.88 \text{ in}$$

$$\delta_{41} = 27.37 \text{ in} \frac{F_{41}}{w_4} = 27.37 \text{ in} \frac{108.6 \text{ kips}}{645 \text{ kips}} = 4.61 \text{ in}$$

Similarly, mode 2 inelastic displacement is calculated as follows:

$$\delta_{x2} = \frac{C_d}{I_e} \delta_{xe2} = \frac{5.5}{1} \left(\frac{386.4 \text{ in/s}^2}{4\pi^2} \right) \frac{(0.241 \text{ s})^2 F_{x2}}{w_x} = 3.13 \text{ in} \frac{F_{x2}}{w_x}$$

The inelastic displacements at each level x are then determined as follows:

$$\delta_{12} = 3.13 \text{ in} \frac{F_{12}}{w_1} = 3.13 \text{ in} \frac{40.4 \text{ kips}}{789 \text{ kips}} = 0.16 \text{ in}$$

$$\delta_{22} = 3.13 \text{ in} \frac{F_{22}}{w_2} = 3.13 \text{ in} \frac{60.8 \text{ kips}}{789 \text{ kips}} = 0.24 \text{ in}$$

$$\delta_{32} = 3.13 \text{ in} \frac{F_{32}}{w_3} = 3.13 \text{ in} \frac{9.0 \text{ kips}}{789 \text{ kips}} = 0.036 \text{ in}$$

$$\delta_{42} = 3.13 \text{ in} \frac{F_{42}}{w_4} = 3.13 \text{ in} \frac{-48 \text{ kips}}{645 \text{ kips}} = -0.23 \text{ in}$$

b. Combine inelastic displacements using SRSS (Equation 7.68) to obtain total inelastic displacements $\delta_x = \sqrt{\Sigma \delta_{xi}^2}$.

$$\delta_1 = \sqrt{(1.02 \text{ in})^2 + (0.16 \text{ in})^2} = 1.03 \text{ in}$$

$$\delta_2 = \sqrt{(2.44 \text{ in})^2 + (0.24 \text{ in})^2} = 2.45 \text{ in}$$

$$\delta_3 = \sqrt{(3.88 \text{ in})^2 + (0.036 \text{ in})^2} = 3.88 \text{ in}$$

$$\delta_4 = \sqrt{(4.61 \text{ in})^2 + (-0.23 \text{ in})^2} = 4.61 \text{ in}$$

c. Determine drifts at each level using Equation 8.19. The story drifts associated with mode 1 are determined below:

$$\Delta_{11} = \delta_{11} - \delta_{1-1,1} = \delta_{11} - \delta_{01} = 1.02 \text{ in} - 0 = 1.02 \text{ in}$$

$$\Delta_{21} = \delta_{21} - \delta_{2-1,1} = \delta_{21} - \delta_{11} = 2.44\,\text{in} - 1.02\,\text{in} = 1.42\,\text{in}$$

$$\Delta_{31} = \delta_{31} - \delta_{3-1,1} = \delta_{31} - \delta_{21} = 3.88\,\text{in} - 2.44\,\text{in} = 1.44\,\text{in}$$

$$\Delta_{41} = \delta_{41} - \delta_{4-1,1} = \delta_{41} - \delta_{31} = 4.61\,\text{in} - 3.88\,\text{in} = 0.73\,\text{in}$$

Similarly, the story drifts associated with mode 2 are determined as follows:

$$\Delta_{12} = \delta_{12} - \delta_{1-1,2} = \delta_{12} - \delta_{02} = 0.16\,\text{in} - 0 = 0.16\,\text{in}$$

$$\Delta_{22} = \delta_{22} - \delta_{2-1,2} = \delta_{22} - \delta_{12} = 0.24\,\text{in} - 0.16\,\text{in} = 0.08\,\text{in}$$

$$\Delta_{32} = \delta_{32} - \delta_{3-1,2} = \delta_{32} - \delta_{22} = 0.036\,\text{in} - 0.24\,\text{in} = -0.21\,\text{in}$$

$$\Delta_{42} = \delta_{42} - \delta_{4-1,2} = \delta_{42} - \delta_{32} = -0.23\,\text{in} - 0.036\,\text{in} = -0.27\,\text{in}$$

d. Combine drifts using SRSS (Equation 7.68) to obtain total drifts $\Delta_x = \sqrt{\Sigma \Delta_{xi}^2}$.

$$\Delta_1 = \sqrt{(1.02\,\text{in})^2 + (0.16\,\text{in})^2} = 1.03\,\text{in}$$

$$\Delta_2 = \sqrt{(1.42\,\text{in})^2 + (0.08\,\text{in})^2} = 1.42\,\text{in}$$

$$\Delta_3 = \sqrt{(1.44\,\text{in})^2 + (-0.21\,\text{in})^2} = 1.45\,\text{in}$$

$$\Delta_4 = \sqrt{(0.73\,\text{in})^2 + (-0.27\,\text{in})^2} = 0.78\,\text{in}$$

e. Check allowable story drifts per ASCE-7, Section 12.12.1 and Table 12.12-1. For buildings in Occupancy Category II, this table gives an allowable drift (for four stories or less structures with interior walls, partitions, etc. designed to accommodate story drifts) of

$$\Delta_a = 0.025\,h_{sx}$$

where:
h_{sx} is the story height for level x. In this case each level is 12 ft high; thus,

$$\Delta_a = 0.025\,h_{sx} = 0.025(12\,\text{ft})(12\,\text{in/ft}) = 3.6\,\text{in}$$

Therefore, none of the story drifts exceed the allowable drift!

8.7 INTRODUCTION TO SEISMIC RESPONSE HISTORY PROCEDURES

The code allows both linear and nonlinear seismic response history analyses following specifications in ASCE-7, Chapter 16. Here, we briefly introduce the linear analysis procedure; details of the nonlinear analysis are beyond the scope of this textbook. In both cases, we first develop a mathematical model of the structural system, which typically involves a finite element model. After creating the mathematical model (which can be two- or three-dimensional), we must select a suite of three or more ground motion acceleration histories that are compatible with the design response spectrum for the site. Appropriate ground motion records include those that have magnitudes, fault distance, and source mechanisms consistent with those that control the maximum considered ground motion at the site. Simulated ground motion records may be used when recorded ground motions are not available. In order to include the effects of ductility and adjust the design response for various hazard levels, the response parameters should be divided by R/I_e.

The mathematical model of the structure is analyzed by directly solving the coupled equations of motion discussed in Chapter 7, Equation 7.45, but including damping,

$$[m]\{\ddot{u}\} + [c]\{\dot{u}\} + [k]\{u\} = -[m]\{1\}\ddot{u}_g \qquad [8.28]$$

These equations can be solved using one of two procedures: direct analysis or modal analysis. Direct analysis requires the numerical integration of the coupled equations of motion using one of several time-stepping techniques, such as Newmark's method. This requires the explicit formation of the system damping matrix [c], which is not required in a modal analysis. The numerical integration of MDOF equations is similar to the process for SDOF systems covered in Sections 4.4 and 4.5, but can be much more complex depending on the number of degrees of freedom (see Villaverde 2009 for more details). The modal analysis, on the other hand, uses mode shapes to transform the coupled MDOF system of equations into a series of SDOF equations (with modal amplitudes as the unknowns), one for each mode—this is similar to an MRS analysis. After the amplitudes are determined, they are transformed back to the nodal displacements using mode shapes. The two approaches produce equal results; however, the modal response history analysis method enjoys several advantages over the direct integration method, including the fact that each SDOF equation can be solved using the methods covered in Sections 4.4 and 4.5, the explicit formation of the system damping matrix [c] is not required, and accurate results may be obtained using a limited number of modes (Villaverde 2009).

Once the deformation response history is obtained, a complete analysis can be performed following the approach presented in Section 3.3. That is, after determining the maximum dynamic displacement from a dynamic excitation, we can conduct a static structural analysis to determine element forces (bending moment, shear force, and axial force) and stresses needed for design; no additional dynamic analysis is necessary.

PROBLEMS

8.1 A column of a steel-braced frame in a single-story building carries the following external loading: dead load, $P_D = 35$ kips, roof live load, $P_{Lr} = 15$ kips, and seismic load (from horizontal effects), $Q_E = \pm15$ kips. Also, $S_{DS} = 1.25$, $\Omega_o = 2.5$, and $\rho = 1.3$. Determine the maximum and minimum axial forces in the column using ASD load combinations. Repeat the problem accounting for overstrength.

8.2 Given the following frame and loading (moment and axial load) shown, determine the maximum and minimum design loads in column C5 using LRFD load combinations. $S_{DS} = 0.65$ and redundancy factor $\rho = 1.0$.

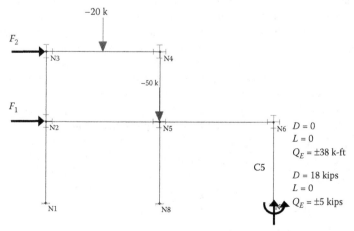

8.3 Given the following single-story office building frame and loading, determine the maximum and minimum design loads in the brace member A, and maximum design load in column B using ASD load combinations. $S_{DS} = 0.65$ and redundancy factor $\rho = 1.3$.

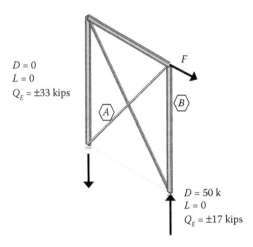

8.4 Use ASCE-7, Table 12.2-1, to determine seismic force-resisting system parameters and restrictions for a steel special concentrically braced frame (SSCBF). Also, list an advantage and a disadvantage for SSCBF systems.

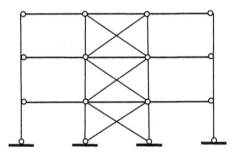

8.5 Given $S_s = 1.82$ and $S_1 = 0.68$, and assuming a stiff soil profile, determine the site coefficients F_a and F_v, the maximum spectral accelerations S_{MS} and S_{M1}, and the design spectral accelerations S_{DS} and S_{D1}.

8.6 Determine the spectral response acceleration parameters and draw the response spectrum for the Golden 1 Center Arena site, located at 500 David J Stern Walk, Sacramento, California. Assume that the site (soil) class is D.

8.7 Draw the design response spectrum for a site where there is very dense soil (Site Class C). The geotechnical engineer has provided the mapped acceleration parameters ($S_S = 1.25$ and $S_1 = 0.52$) and the long-period $T_L = 16$ s.

8.8 A seismic vulnerability assessment of a two-story ordinary building located in Stockton, California, is being performed. If the building's first story is 14 ft high and the second story is 12 ft high, compute the approximate structural period T_a, the static design base shear V, the lateral forces at each level F_x, and the overturning moment at the base. The proposed structural system is a steel ordinary moment frame with an estimated weight of 50 kips on the first level and 20 kips on the roof. The geotechnical engineer has provided the mapped design acceleration parameters ($S_{DS} = 0.672$ and $S_{D1} = 0.362$).

8.9 Given a two-story, 25-foot average height, wood frame single-family residence located in Stockton, California, compute the design base shear. The geotechnical engineer estimates that the house will be on Site Class D, with mapped acceleration parameters of $S_S = 1.75$ and $S_1 = 0.90$. The building in plan view is regular. The structural system being proposed is a wood structural panel shearwall. The estimated weight on level 1 is 83 kips and level 2 is 52 kips.

8.10 Given a six-story office building frame with equal story heights of 12 ft located in Brookside (Stockton, California), compute the design earthquake loads acting on each floor. The building in plan view is square. The structural system being proposed is a steel special concentrically braced frame with an estimated weight on each floor of 100 kips and 80 kips on the roof. The geotechnical engineering report has classified the site soil class as D.

8.11 First, sketch the design response spectrum using $S_{DS} = 0.82$ and $S_{D1} = 0.47$. Then, use the equivalent lateral force procedure to compute the approximate structural period T_a, the design base shear V, the lateral forces at each level F_x, and the overturning moment at the base. The structural system for the office building is a steel ordinary moment frame.

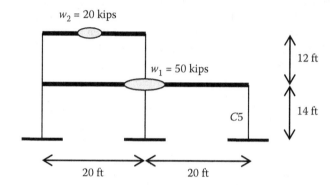

8.12 A four-story special moment-resisting steel frame with equal story heights of 15 ft has a period of 0.7 s, occupancy category of III, and elastic displacement of 0.7 in. What is the maximum inelastic displacement (δ_x) for the first story?

8.13 A three-story hospital building with reinforced concrete special moment frames has floor heights of $h_1 = 14$ ft, $h_2 = 12$ ft, and $h_3 = 10$ ft. The elastic displacements at each story are $\delta_{1e} = 0.8$ in, $\delta_{2e} = 1.45$ in, and $\delta_{3e} = 2.15$ in. $S_{DS} = 0.82$ and $S_{D1} = 0.47$, SDC of D, and period, $T = 0.75$ s. Find the design story drifts and check to ensure that they are within the allowable drifts.

8.14 Given a four-story County Jail with 12-foot story heights and steel special concentrically braced frames, determine design story drifts and verify that they are within allowable drift limits. The equivalent lateral force procedure analysis results are shown below.

Level	w_x (kips)	F_x (kips)	δ_{xe} (in)
4 (roof)	100	25	1.23
3	120	20	0.80
2	120	13	0.45
1	130	8	0.20

8.15 Use the equivalent lateral force procedure to compute the approximate structural period T_a, the base shear V, the lateral forces at each level F_x, the overturning moment at the base, and the inelastic story drifts given the elastic displacements shown. Also, determine if the drifts are within the allowable limits. The structural system is a special moment reinforced concrete frame at a site with $S_{DS} = 0.82$ and $S_{D1} = 0.47$.

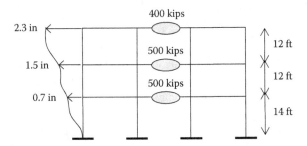

8.16 A three-story office building with a special moment-resisting frame has floor heights of $h_1 = 15$ ft, $h_2 = 12$ ft, and $h_3 = 12$ ft. The building has an estimated weight on each floor of 850 kips and 700 kips on the roof. $S_{DS} = 0.81$, $S_{D1} = 0.50$ and seismic design category D. A preliminary analysis yielded mode shapes and periods shown. First, use the ELFP method to determine the seismic base shear, and then use the MRS analysis method to determine effective weight for each mode, story forces for each mode, and the maximum story shear forces by SRSS.

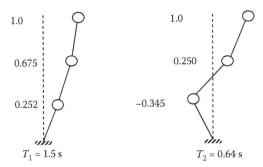

8.17 Given a three-story building and corresponding eigensolution (normalized eigenvectors and eigenvalues), determine the periods, participation factors, and effective modal weights for each mode shape. Also, use the SRSS method to determine the maximum elastic-based shear using the fewest number of modes such that they contribute 90% of the effective modal weight, per ASCE-7 MRS analysis.

$$[\Phi] = \begin{bmatrix} 1.00 & 1.00 & 1.00 \\ 0.64 & -0.60 & -2.57 \\ 0.30 & -0.68 & 2.47 \end{bmatrix}$$

$$[\lambda] = \begin{bmatrix} 21 & 0 & 0 \\ 0 & 96.6 & 0 \\ 0 & 0 & 221.4 \end{bmatrix} s^{-2}$$

$$\text{Assume } \{S_a\} = \begin{Bmatrix} 88.8 \\ 187.2 \\ 272.4 \end{Bmatrix} \text{ in/s}^2$$

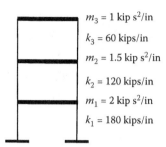

$m_3 = 1$ kip s^2/in

$k_3 = 60$ kips/in

$m_2 = 1.5$ kip s^2/in

$k_2 = 120$ kips/in

$m_1 = 2$ kip s^2/in

$k_1 = 180$ kips/in

8.18 Given the following eigensolution for a three-story building with floor weights of 1800 kips for the first floor and roof, and 1200 kips for the second floor:

$$[\Phi] = \begin{bmatrix} 2.86 & -0.657 & 0.387 \\ 1.95 & 0.725 & -1.61 \\ 1.00 & 1.00 & 1.00 \end{bmatrix} \quad \text{and} \quad \{\omega\} = \begin{Bmatrix} 15.1 \\ 38.5 \\ 61.7 \end{Bmatrix} \text{ rad/s, which yield } \{S_a\} = \begin{Bmatrix} 377 \\ 346 \\ 308 \end{Bmatrix} \text{ in/s}^2$$

Determine the participation factors for each mode, effective weight for each mode, base shears for each mode using the given design spectral accelerations, and the maximum base shear using the SRSS method.

8.19 Given the following six-story office building frame, ground motion parameters, and mass and stiffness matrices, use MATLAB to compute the displacements, design earthquake loads acting at each floor, base shear, and overturning moment using the MRS analysis. Also, compute the effective modal masses for the six modes and accumulated mass percentage.

Ground motion parameters:
   Site Class $= D$
   $S_{DS} = 0.323$
   $S_{D1} = 0.186$

Building frame parameters (steel intermediate moment frame):
   Risk category $=$ II (office building)
   SDC $= C$
   $R = 4.5$
   $C_d = 4$

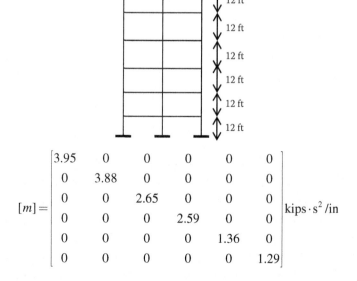

12 ft

12 ft

12 ft

12 ft

12 ft

12 ft

$$[m] = \begin{bmatrix} 3.95 & 0 & 0 & 0 & 0 & 0 \\ 0 & 3.88 & 0 & 0 & 0 & 0 \\ 0 & 0 & 2.65 & 0 & 0 & 0 \\ 0 & 0 & 0 & 2.59 & 0 & 0 \\ 0 & 0 & 0 & 0 & 1.36 & 0 \\ 0 & 0 & 0 & 0 & 0 & 1.29 \end{bmatrix} \text{ kips} \cdot \text{s}^2/\text{in}$$

$$[k] = \begin{bmatrix} 2260.0 & -1374.7 & 198.5 & -28.4 & 2.0 & -0.2 \\ -1374.7 & 1822.9 & -896.6 & 191.6 & 12.9 & 1.44 \\ 198.5 & -896.6 & 1352.6 & -714.8 & 91.9 & -10.5 \\ -28.4 & 191.6 & -714.8 & 885.8 & -410.2 & 73.0 \\ 2.0 & -12.9 & 91.9 & -410.2 & 521.9 & -192.5 \\ -0.2 & 1.44 & -10.5 & 73.0 & -192.5 & 128.7 \end{bmatrix} \text{kips/in}$$

REFERENCES

American Society of Civil Engineers, ASCE/SEI, *Minimum Design Loads for Buildings and Other Structures*, Reston, VA, 2010.
Villaverde, R., *Fundamental Concepts of Earthquake Engineering*, CRC Press, Boca Raton, FL, 2009.

Index

A

Accelerations, 34, 37, 80
Aftershocks, 20
Allowable seismic force-resisting system, 206–207
Allowable stress design (ASD), 203–206
American Society of Civil Engineers (ASCE), 15
Ancash earthquake, 5
ASCE-7, 42, 203, 215, 219
    earthquake loads based on ASCE-7 equivalent lateral
        force procedure, 220–229
    ELFP analysis steps, 220–225
    MRS analysis steps, 229
ASCE, *see* American Society of Civil Engineers
ASD, *see* Allowable stress design
Ashlar masonry, 15
Attenuation, 29
    relations, 138

B

Base shear
    using participation factors and response spectra,
        157–159, 163–165, 182–187
    by time-dependent forces and support excitations,
        83–85
Bending moment, 53–54
Body waves, 21, 22
Braced frames, 51, 53–56
Bracketed duration, 30–31

C

California Strong Motion Instrumentation Program
    (CSMIP), 33
Cantilever shear wall, 57
Central difference method, 117–120
Circular frequency, 42
City of Los Angeles building code (1943), 16
Collapse earthquakes, 9
    comparison of building collapse and natural
        earthquakes, 10
Complementary solutions, 62–63, 74
Complete quadratic combination methods
    (CQC methods), 169, 185
Compliant systems, 33
Contemporary design spectral analysis, 138
Continental lithosphere, 11
Continuous system, 165–167; *see also* Discrete system
    deflections, base shear, and moments, 163–165
    equation of motion, 162–163
    generalized participation factor, frequency, and
        natural period, 163
    generalized SDOF, 162
conv function, 102
Convolution integral, 91, 114
CQC methods, *see* Complete quadratic combination
    methods

Critical damping coefficient, 68
Critically damped system, 67
CSMIP, *see* California Strong Motion Instrumentation
    Program

D

D'Alembert's principle, 61, 65–66, 152, 171
Damped single-degree-of-freedom system, 64; *see also*
    Undamped single-degree-of-freedom system
    free vibration response of damped systems, 65–71
    structural damping, 71–73
    time-dependent forced damped vibration response,
        73–78
    time-dependent support accelerations, 78–82
Damped systems, free vibration response of, 65–71
Damping, 32, 91
    coefficient, 66
    ratio, 71
Dead loads (DL), 42, 204, 206, 221
Decoupling process, 176–177
Deflections, 157–159, 163–165, 182–187, 212
Deformation response factor, 63, 74
Degree-of-freedom (DOF), 37–38, 171
Design respsponse spectrum, 213–218
Detailing, 211
Deterministic approach, 34
Diagonal matrix, 172
Dip-slip, 20
Dip angle, 19
Direct integration methods, 114
    central difference method, 117–120
    Newmark's Beta method for linear systems, 120–121
    Nigam–Jennings algorithm, 114–117, 118
Discrete system, 152; *see also* Continuous system
    deflections, base shear, and moments, 157–159
    equation of motion, 152–156
    example, 159–161
    generalized participation factor, frequency, and
        natural period, 157
Displacements, 37
DL, *see* Dead loads
DOF, *see* Degree-of-freedom
Dry-stone walls, 15
Ductility, 33, 71
Duhamel's integral, 91, 95
Dynamic analysis, 37, 83
Dynamic magnification factor, *see* Deformation response
    factor

E

Earthquake, 1, 37
    accelerogram, 123
    Ancash earthquake, 5
    charactereristics, 29–31
    development of mitigation strategies, 14–17
    earthquake-resistant design, 14

Printed in the United States
by Baker & Taylor Publisher Services